The Golden Mouse

The Golden Mouse

Ecology and Conservation

Edited by

Gary W. Barrett
University of Georgia, Athens, GA, USA

George A. Feldhamer
Southern Illinois University, Carbondale, IL, USA

 Springer

Editors
Dr. Gary W. Barrett
University of Georgia
Eugene P. Odum School of Ecology
Athens, GA 30602-2202
gbarrett@uga.edu

Dr. George A. Feldhamer
Southern Illinois University
Department of Zoology
Carbondale, IL 62901-6511
feldhamer@zoology.siu.edu

ISBN: 978-0-387-33665-7   e-ISBN: 978-0-387-33666-4

Library of Congress Control Number: 2007924492

Printed on acid-free paper

9 8 7 6 5 4 3 2 1

springer.com

To Kagen Michael and Brittnee Paige
May you always love wildlife and appreciate
the search for Nature's treasures

GWB

To Carla, Andy, and Carrie
As always

GAF

Foreword

The golden mouse (*Ochrotomys nuttalli*) of the southeastern United States and a diagnostic species of the Austroriparian biotic province exhibits a number of distinctive features in its morphology, physiology, behavior, life history, and ecology. The bright ochraceous color of the upper parts, from which its generic name is derived, contrasts sharply with the more muted pelage coloration of other species of native rats and mice of eastern North America. More fundamental morphological features, including skull and dental characteristics, anatomy of the male reproductive tract, and sperm morphology, also clearly differentiate *Ochrotomys* from other rodent genera. With the exception of the tree and flying squirrels, the golden mouse also is arguably the most arboreal small mammal in eastern North America. Although it might nest on the ground beneath the leaf litter, it frequently constructs conspicuous globular aboveground nests in shrubs or trees, often in hanging vines, such as honeysuckle and greenbrier. Its semi-aboreal habits are reflected in such features of its morphology as a semi-prehensile tail, well-developed plantar tubercles, and strong abdominal musculature. A low basal metabolic rate and tendency to become lethargic at high environmental temperatures are presumably physiological adaptations to reduce heat stress experienced from inhabiting arboreal nests exposed to high summer temperatures. Certain behavioral features such as a tendency of adults to "freeze" when disturbed and the relatively rapid growth and development and low exploratory tendencies of the young also appear to be adaptations for arboreal activity.

Since it was first made known to science by Harlan in 1832, the golden mouse has been the subject of numerous studies dealing with various aspects of its taxonomy, biology, ecology, and behavior. The present volume represents the first attempt to compile and synthesize this substantial body of information. Hopefully, it will not only provide a valuable summary of current knowledge of the golden mouse, but will also reveal what we do not know about the species and thus serve as a stimulus for further research on this distinctive and attractive small mammal.

James N. Layne

Preface

When we hear the familiar phrase *charismatic mammalian megafauna*, we immediately envision large, powerful carnivores like lions and grizzly bears, or sleek graceful ungulates like deer and antelope. However, we rarely, if ever, hear about charismatic mammalian *microfauna* such as rodents. In fact, most people consider small, cryptic rodents as nothing more than vectors of disease, crop depredators, or prey for larger and much more interesting mammals. Yet, many rodent species serve as critical models for medical or ecological research, are valuable as furbearers and sustenance, are important in the pet trade, or possess novel and compelling life history characteristics. In terms of fascinating life history, the subject of this volume—the golden mouse (*Ochrotomys nuttalli*)—has few equals. For example, it exhibits unique patterns of behavior related to bioenergetics, nest building, coexistence with sympatric species of small mammals, and potential longevity. A fairly rare species throughout its geographic range, the golden mouse usually inhabits areas with very thick, dense understory vegetation where it builds softball-sized arboreal nests, as well as ground nests more typical of woodland mice. During certain times of the year, several golden mice might communally occupy large arboreal nests termed shelter/communal nests. In addition to their intriguing life history characteristics, the strikingly radiant golden color of their fur makes them a particularly intriguing and appealing component of the small mammal fauna.

In this volume, we bring together zoologists, ecologists, behaviorists, parasitologists, artists, and other authorities to contribute their expertise to an investigation and a better understanding of the golden mouse. Each author brings his or her experience and insights from being directly involved with ongoing research related to the golden mouse and related species of small mammals. We have attempted to produce a concise, scholarly work based on past and current research that will be useful to students and professionals in mammalogy, ecology, and wildlife biology, as well as general readers interested in natural history. We use the golden mouse as the focal species to explore conceptual issues in ecology across levels of organization (individual, population, community, ecosystem, and landscape), integrating reductionist and holistic ecological science.

Chapter 1 provides an overview of the golden mouse explaining why a levels-of-organization approach is used to organize information and to suggest future investigations in order to better understand this unique species. Following an historical perspective, examples are presented to introduce early chapters in the book based on the levels-of-organization concept. Latter chapters focus on ecological processes (e.g., regulation, energetics, and behavior) that transcend these levels of organization.

At the individual level (Chapter 2), we discuss natural history, taxonomy, evolution, and systematics of the golden mouse. We then move to the population level (Chapter 3), where growth, population dynamics, and population genetics are described for this species. At the community level (Chapter 4), topics such as coexistence, competitive interactions, and the benefits of semiarboreal living are discussed. Natural and anthropogenic perturbations, secondary succession, and the impact of various forestry practices on ecosystem dynamics that affect golden mice are discussed in Chapter 5. At the landscape level (Chapter 6), we discuss riparian habitats as possible landscape corridors, landscape fragmentation, and patch quality relative to what is known about golden mice and identify questions yet to be addressed.

Chapters 7–10 discuss such transcending processes as rarity, energetics, behavior, and parasitism. For example, the status of the golden mouse as a relatively rare species and conservation and management practices related to this species are outlined in Chapter 7. A discussion describing why the golden mouse represents a model species to better understand mechanisms of energetic efficiency is presented in Chapter 8. Because of its diverse patterns of nest construction and unusual nesting behavior, the authors in Chapter 9 suggest that the golden mouse be considered an ecological (*oikos*) engineer. Chapter 10 summarizes the ectoparasites found on golden mice and the relationship of golden mice to vector-borne diseases.

Finally, the authors place and discuss golden mice within the larger perspective of landscape aesthetics (Chapter 11). In Chapter 12 the editors of this volume present future investigative challenges and outline important questions yet to be addressed. We hope by demonstrating how a relatively small mammal species can be investigated using this approach—unlike a taxon such as *Peromyscus*—will help to define future areas of research and will promote future integrative studies across all levels of organization. We also hope that undergraduates, graduate students, working professionals, and interested laypersons will agree that the golden mouse is a worthy standard bearer for, and prime example of, the most charismatic of mammalian microfauna.

Gary W. Barrett
George A. Feldhamer

Acknowledgments

We thank Janet Slobodien, Editor, Life Sciences, Springer for her support of the book since its conception. We also thank Felix Portnoy, Production Editor, Springer for his coordination of the production of this book. We express our appreciation to Joseph Piliero, Design Manager, Springer, for the evocative cover design of this book. Naturally, special thanks are extended to the contributors of each of the following chapters. Special thanks are also due to Andrea Zlabis, Brookfield Zoological Park, Chicago, Illinois for sharing her experience and for providing photographs of the golden mouse exhibit in "The Swamp" area located at the Zoological Park; to Joy Richardson, Barrow County School System, Georgia, for directing our attention to the books on Poppy, the golden mouse; and to Thomas Luhring, Anita Morzillo, Tim Carter, Terry Barrett, and James Layne for their photographs that appear in this book. We are grateful to James Layne for providing the Foreword for the book. We are especially grateful to Terry Barrett for her time, effort, and knowledge regarding numerous editorial suggestions and the selection of photographs and graphics used throughout the text. The invaluable assistance of Lisa Russell, Environmental Studies Program, Southern Illinois University Carbondale (SIUC), in several phases of this project is gratefully acknowledged. Thanks are also extended to Jonathan Gray, Department of Speech Communication at SIUC, and James Layne, Archbold Biological Station, Lake Placid, Florida, for their contributions as reviewers. We acknowledge Margaret Adamic, Contracts Administrator Publishing, Corporate Legal, Disney Publishing Worldwide, for archival research pertaining to and assistance in obtaining permissions for the ©Disney Enterprises, Inc. images shown in this volume.

We are indebted to numerous students (both undergraduate and graduate) from Miami University of Ohio and the Eugene P. Odum School of Ecology, University of Georgia, who over the past several decades have provided assistance in the field. Many of these individuals have gone on to become professionals in fields such as law, medicine, business, and academia. This list includes, but is not limited to, Molly Anderson, Jill Auburn, Jennifer Blesh, Susan Brewer, Megan Casey, Jennifer Chastant, Cory Christopher, Daniel Crawford, Scott Davis, Brett Dietz, Janet Ford, Tara Gancos, Laura Gibbes, Patricia Gregory, Lauren Hall, Steve Harper, Chad Jennison, Melissa Jewell, Ryan Klee, Barbara Knuth, Chris Lucia, Thomas Luhring, Anika Mahoney, Mark Maly, Kelli Meek, Maura O'Malley, Cayce

Payton, John Peles, Bill Peterjohn, Kevin Postma, Alison Pruett, Luis Rodas, Jeff Ryan, Chris Schmidt, Steve Shivers, Scott Springer, Matthew Shuman, Ryan Stander, Karen Stueck, Christopher Williams, Michelle Williams, and Sabrina Willis. Many of these individuals conducted independent research on the golden mouse as undergraduates and had their findings published in professional peer-reviewed journals. Likewise, we drew on the work of several SIUC graduate students whose thesis research involved aspects of the golden mouse, including: Dean Corgiat, Chelsea DeBay, Kathy Furtak-Maycroft, Anita Morzillo, and David Wagner. Witnessing the academic growth and success of these individuals is one of the highlights of our careers as educators.

Contents

Section 3   Transcending Processes

Section 4   New Perspectives and Future Challenges

Contributors

GARY W. BARRETT
Eugene P. Odum School of Ecology, University of Georgia, Athens, GA 30602

TERRY L. BARRETT
120 Riverbottom Circle, Athens, GA 30606

GUY N. CAMERON
Department of Biological Sciences, University of Cincinnati, Cincinnati, OH 45221

DONALD W. LINZEY
Department of Biology, Wytheville Community College, 1000 East Main Street, Wytheville, VA 24382

THOMAS M. LUHRING
Eugene P. Odum School of Ecology, University of Georgia, Athens, GA 30602

ANITA T. MORZILLO
US EPA, 200 SW 35th Street, Corvallis, OR 97333

JOHN D. PELES
Penn State University-McKeesport, 207 Ostermayer Laboratory, McKeesport, PA 15132

CORY C. CHRISTOPHER
Department of Biological Sciences, University of Cincinnati, Cincinnati, OH 45221

LANCE A. DURDEN
Department of Biology, Georgia Southern University, Statesboro, GA 30460

GEORGE A. FELDHAMER
Department of Zoology, Southern Illinois University, Carbondale, IL 62901

JAMES N. LAYNE
Archbold Biological Station, Lake Placid, FL 33852

ROBERT K. ROSE
Department of Biological Sciences, Old Dominion University, Norfolk, VA 23529

STEVEN W. SEAGLE
Department of Biology, Appalachian State University, 572 Rivers Street, Boone, NC 28606

JERRY O. WOLFF
Department of Biological Sciences, St. Cloud State University, 720 Fourth Avenue South, St. Cloud, MN 56301

About the Editors

Gary W. Barrett is Odum Chair of Ecology in the Eugene P. Odum School of Ecology at the University of Georgia. Until 1994, he was Distinguished Professor of Ecology at Miami University, Oxford, Ohio, where he founded the Institute of Environmental Sciences and the Ecology Research Center. From 1994 through 1996 he was the director of the Eugene P. Odum School of Ecology at the University of Georgia. He is the author or coauthor of 6 books and over 170 publications in professional journals. He was Ecology Program Director with the National Science Foundation from 1981 to 1983 and has served on or chaired numerous committees within the American Society of Mammalogists, American Institute of Biological Sciences, International Association for Landscape Ecology, International Association for Ecology, and National Research Council of the National Academy of Sciences. Barrett is a fellow of the American Association for the Advancement of Science (1990). He served as chair of the Applied Ecology Section of the Ecological Society of America (1985–1987), president of the United States Regional Association of the International Association for Landscape Ecology (1988–1990), president of the Association for Ecosystem Research Centers (1995–1996), and president of the American Institute of Biological Sciences (1998). He received the AIBS Presidential Citation Award in 2000 in recognition of leadership and contributions to the biological sciences and the prestigious Distinguished Landscape Ecologist Award in 2001 from the United States Regional Association of the International Association for Landscape Ecology. Barrett was the recipient of the Excellence in Undergraduate Mentoring Award at the University of Georgia in 2005.

 *George A. Feldhamer* is Professor of Zoology and Director of the Environmental Studies Program at Southern Illinois University at Carbondale (SIUC). His research has focused exclusively on mammalian populations, ecology, and management; biology of introduced cervids; and threatened and endangered species, with funding from state and federal management agencies. He has published in numerous professional journals. He was associate editor for forest biology and ecology for the *Journal of Forest Research* (1997–2004) and the *Wildlife Society Bulletin* (1993–1995). He is senior editor of *Wild Mammals of North America: Biology, Management, and Conservation* (2003) and coauthor of *Mammals of the National Parks: Conserving America's Wildlife and Parklands* (2005), both published by John Hopkins University Press. He is senior author of the textbook *Mammalogy: Adaptation, Diversity, Ecology* (1999; 2003) published by McGraw-Hill. He has 30 years of experience in teaching upper-division mammalogy courses, and a game mammal management course at SIUC and the University of Maryland. In 2000, Feldhamer was named Outstanding Teacher in the College of Science at SIUC.

Section 1
Introduction

1
The Golden Mouse:
A Levels-of-Organization Perspective

Gary W. Barrett

Who else had known and admired golden mice? Theodore Roosevelt, twenty-sixth president of the United States, knew of them. "As a boy I worked in the museum and . . . remember skinning some rather reddish white-footed mice I thought were golden mice, and was disappointed to find they were not." (Terres 1966:98)

A Personal History Perspective

I vividly recall the first time I observed a golden mouse (*Ochrotomys nuttalli*) in the field. It was the summer in 1965 while conducting my doctoral dissertation research at the University of Georgia (UGA). My dissertation research was the first major investigation conducted at HorseShoe Bend (HSB) Experimental Site located in Clarke County, latitude 33°57′N and longitude 83°23′W, near Athens, Georgia. HSB is a 35-acre (14.1-ha) research site created by a meander of the North Oconee River (see Barrett 1968, Blesh and Williams 2003, Hendrix 1997 for detailed descriptions of the site). This ecosystem-level investigation focused on the effects of a carbamate insecticide (Sevin) on small mammal populations in semi-enclosed grassland ecosystems later published in *Ecology* (Barrett 1968). At HSB, the undergrowth outside of the 0.4-ha enclosures and along the North Oconee River contained an abundance of Chinese privet (*Ligustrum sinense*), Japanese honeysuckle (*Lonicera japonica*), and greenbrier (*Smilax* sp.) within a bottomland forest community. One afternoon during the 1960s, while hiking through this mixed hardwood and thicket-type plant community, I came upon a globular nest. I had never seen a nest like this during my childhood while living on a farm in southern Indiana. I assumed it was a bird nest of which I was not yet familiar. Out of curiosity, I touched the nest with a stick and out from the nest appeared the most beautiful small mammal that I had ever observed. This docile animal stopped on a limb in the sun to groom. Its rich golden pelage remains vivid in my mind to this day. I returned to the nest site the next day with a pair of long forceps to live capture (by its tail) and identify this beautiful small mammal species. I found it interesting to observe a small mammal residing in bushes, rather than in open fields, such as the deer mice (*Peromyscus maniculatus*) that I had livetrapped in a red clover (*Trifolium pratense*) field while conducting research for my Master's thesis at Marquette University (Barrett and Darnell 1967). This individual was identified as a golden mouse and that experience, unrelated to my dissertation research at the time, was my initial introduction to *O. nuttalli*. It was also at that time that I read an article in the March–April issue of *Audubon Magazine* entitled "Search for the Golden Mouse" (Terres 1966), which intrigued me. These circumstances served as an early incentive for a continuing interest that would result many years later in this book.

Following the award of my Ph.D. at the University of Georgia in 1967, and after one year on the faculty at Drake University, I joined the faculty at Miami University in Oxford, Ohio in 1968. I soon recognized from geographical distribution maps (see Figure 2.1 in Chapter 2 of this volume) that *O. nuttalli* did not occur in southern Ohio but was present in Kentucky. It appears that the Ohio River has served as a natural barrier and boundary regarding the northern geographic range of this species in that area.

During early December 1973, a group of students accompanied me on what would become a pilgrimage to Lexington, Kentucky to determine if we could locate and live capture golden mice for a bioenergetics feeding study (see Stueck et al. 1977 for details). We met for breakfast with the late Roger W. Barbour to discuss sites in Kentucky where we might capture golden mice. Dr. Barbour, along with William H. Davis, was the author of *The Mammals of Kentucky* (Barbour and Davis 1974). Dr. Barbour informed us, following several muffins and a cup of hot chocolate, that our best bet was to explore an area where he discovered a population of golden mice several years earlier. This area was located in a box canyon near Big Hill, Madison County, Kentucky. Big Hill is located near Berea College, an institution nationally recognized for its high standards of education for students residing in the Appalachian Mountains of Kentucky and nearby states. Interestingly, students from Berea College were living in what I would describe as a large log house at this site. Our trip to Big Hill was very exciting and successful.

For example, six golden mice (four males and two females) were captured in one large communal/shelter nest located in an Eastern red cedar (*Juniperus virginiana*) tree (Stueck et al. 1977). We also learned that it typically took several days for golden mice to enter live traps (live traps set on this 2-day excursion were unsuccessful). Thus, we developed a "new capture method" during future 2-day excursions to central Kentucky and elsewhere. Once we located a globular or shelter nest, we used a stepladder to get an eye-level view of the nest if possible; some nests were too high for this strategy. Fortunately, several nests near ground level did not require a ladder. One individual would climb the ladder and then, very carefully, touch the nest with 12-in (30.5-cm) forceps (specimen forceps, Carolina Science, FR-62-4335). If golden mice were present, an individual would typically appear near the entrance to the nest. Unless one is very careful (patience is a great virtue while collecting golden mice), one or more individuals in the nest might become alarmed and leap to the ground. Those individuals that emerged from the nest, yet stayed in the brush or tree canopy, remained our focus of capture. Persons on the ground also had forceps and observed the movement of golden mice once they exited a nest.

Because golden mice are typically docile (unless unnecessarily frightened once leaving the nest), they most often will move from branch to branch, frequently from tree to tree, using their prehensile-like tail to aid their arboreal patterns of movement (Goodpaster and Hoffmeister 1954, Packard and Garner 1964). Golden mice typically move slowly through the branches or undergrowth, pausing frequently to groom or rest. With a couple of ladders, much patience, and good eyes, eventually investigators will be successful. They reach through the brush and, with long forceps, carefully, but gently, clamp the tail and place the golden mouse into a pillowcase in which the mouse can be handled by the neck to determine its sex. Captured individuals are then placed in plastic cages for transport back to the laboratory. I have hand-captured literally dozens of golden mice in this manner.

Occasionally one will locate a large communal/shelter nest containing several individuals. I have personally removed six to eight individual golden mice from large communal nests on several occasions (Jewell et al. 1991, Springer et al. 1981, Stueck et al. 1977). If one is careful, most individuals (one at a time) can be captured with forceps as described earlier. Because golden mice are docile, they can even be manipulated with the forceps (assuming that they have been staring at you from a hole in the nest) and then captured by the tail as they return into the nest.

Several other points of interest should be mentioned when one goes "hunting (alive) golden mice." On numerous excursions, 10–15 globular nests (Figure 1.1)

FIGURE 1.1. Representative globular nest constructed and used by golden mice (*O. nuttalli*). Nests might contain several golden mice at the same time. Photograph by Thomas Luhring.

will be located before one discovers an active nest. Why would a golden mouse allocate energy to constructing numerous seldom-used globular nests, or at least not used throughout the growing season? Could these "extra" nests be constructed to divert or decrease rates of capture by snakes or avian predators? These unknowns provide fertile areas for future research.

Occasionally, especially during winter, one comes upon a large communal/shelter nest (see Chapter 9 of this volume for details). These communal nests frequently contain from four to eight individuals (Barbour 1942, Dietz and Barrett 1992, Stueck et al. 1977). Goodpaster and Hoffmeister (1954) and Packard and Garner (1964) also reported that golden mice commonly are found grouped in arboreal nests in winter, but they are more solitary during summer. Springer et al. (1981) observed natural groupings ranging from two to six individuals per nest in late November 1978. Dietz and Barrett (1992) hand-collected two groups of four individuals each from the same nest in Madison County, Kentucky in December 1988. I have observed that large communal/shelter nests frequently have an abundance of sticks and small limbs in addition to finely shredded bark, grasses, leaves, and feathers typical of globular nests (Linzey 1968, Linzey and Packard 1977).

Another important observation: When one locates an active nest with several mice and is only able to capture one or two individuals out of four or more present, those that escaped will return to the nest overnight. Thus, an investigator will get a second opportunity to collect the remaining individuals from that particular nest. Some decisions require common sense and an understanding of the sample research site. For example, one should not collect all individuals from a particular nest unless it was previously confirmed that there is an abundant population density at the site. On several occasions, experimental animals were returned and released at the site of capture following nesting behavior or bioenergetic studies (Jewell et al. 1991, Knuth and Barrett 1984, Peles et al. 1995, Springer et al. 1981). It should also be noted, but not recommended unless one desires to collect a kin group (Dietz and Barrett 1992), that all individuals in a single nest can be collected simultaneously by carefully trimming most of the vines from around the nest and then quickly placing the nest, including the mice therein, into a pillowcase (my favorite collecting sack). One must be exceedingly careful while clipping the vines and small limbs not to disturb the inhabitants within the nest. Again, *patience is a virtue*.

A final observation: One cannot set 100 live traps for one day and hope the next day to capture *O. nuttalli*. Even if these traps are set on limbs near active nests, I have failed to live trap even a single golden mouse; thus, the reason for the "forceps live-capture" methodology described earlier. When live traps are set overnight in a golden mouse habitat, one can expect to capture at least a few white-footed mice (*Peromyscus leucopus*), perhaps even an Eastern chipmunk (*Tamias striatus*) or a Southern flying squirrel (*Glaucomys volans*). However, golden mice do not readily enter live traps the first couple of nights (see Feldhamer and Maycroft 1992). Perhaps live traps function as foreign novel stimuli, thus impeding their immediate entrance in a freshly set live trap.

FIGURE 1.2. Wooden platform and chamber containing a Sherman live trap situated 5 ft (1.5 m) high on the trunk of a tree used to estimate population abundance of *O. nuttalli*. Photograph by Thomas Luhring.

We have observed, however, that if one places a live trap on a wooden L-shaped platform mounted 5 ft (1.5 m) high on the trunk of a tree (Figure 1.2), then golden mice, as well as white-footed mice, will be more readily captured (Christopher and Barrett 2006). This height is also their most active use of three-dimensional habitat space (Jennison et al. 2006, Pruett et al. 2002).

In summary, for 26 years (1968–1994) while serving on the faculty at Miami University of Ohio, groups of us made several pilgrimages to Big Hill, Kentucky, collecting and observing *O. nuttalli* in their prime habitat. I returned to this site twice with students from the UGA to monitor the abundance of this small mammal species. Unfortunately, on these expeditions only seven golden mice were observed in 2001 and two golden mice in 2005. To my surprise, however, the log house formerly occupied by graduate students from nearby Berea College was discovered to be a tavern where Ulysses S. Grant stayed along his way to Lexington, Kentucky, during the American Civil War (Figure 1.3). In fact, a small cemetery from the American Civil War is located on this site. Thus, not only has this site been prime habitat for golden mice, but also it is now a designated state historic landmark. One frequently learns varied histories while investigating the natural habitat of their favorite species.

Well, so much for the enjoyment of field observations and methodologies developed for collecting one of the unique small mammals in the southeastern United States. Suffice it to say, golden mice represent one of the most unusual small mammals based on their particular bioenergetics, habitat selection, nesting

FIGURE 1.3. Students on a golden mouse-collecting trip in November, 1980, near Big Hill, Madison County, Kentucky (left to right: Barbara Knuth, Mark Maly, Chris Lucia, and Bill Peterjohn). The log house in the background is actually a tavern where Ulysses S. Grant had stayed during the American Civil War. Photograph by Terry L. Barrett.

behavior, and niche relationships. Next, let us turn to the reasons why this book is organized along levels of biological/ecological organization.

Levels-of-Organization Approach

The levels-of-organization concept ranges from the cellular to the ecosphere levels (Figure 1.4). Figure 1.4 also illustrates how 7 ecological processes transcend and integrate these 11 levels of organization (see Barrett et al. 1997 for details). We elected to organize the chapters of this book based on this concept, focusing on the organism/natural history (Chapter 2), population (Chapter 3), community (Chapter 4), ecosystem (Chapter 5), and landscape (Chapter 6) levels. Figure 1.4 also shows seven transcending processes (behavior, development, diversity, energetics, evolution, integration, and regulation) that transcend all levels of organization. These processes are illustrated throughout the book, with some chapters devoted specifically to processes such as energetics, behavior, and evolutionary relationships. This chapter will provide examples to articulate how the golden mouse functions within each of the above levels-of-organization from organism, population, community, ecosystem, to landscape.

As mentioned earlier, select chapters are devoted to the transcending processes as they relate to the golden mouse. For example, Chapter 7 discusses the importance of rarity regarding the evolution, behavior, and dynamics of this

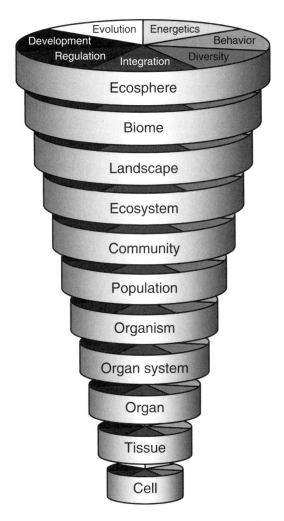

FIGURE 1.4. Model illustrating the levels-of-organization concept. Seven processes transcend and help to integrate levels of organization. Modified after Odum and Barrett 2005, with permission from Brooks/Cole, a division of Thomson Learning.

unique small mammal species. Chapter 8 describes the bioenergetics and energy efficiency of *O. nuttalli* compared to other small mammal species of similar body mass and natural histories. Chapter 9 outlines the unusual nest-building behavior of *O. nuttalli*, illustrating the diversity of nest types constructed and suggesting how these nest types relate to social behavior, ecological energetics, and niche relationships. The authors go so far as to describe the golden mouse as an "ecological *oikos* engineer." Chapter 10 describes the ecology and epidemiology of ectoparasites, bots, and vector-borne diseases associated with golden mice.

Such transcending processes as coevolution, potential role of parasites on golden mouse bioenergetics, and the evolutionary relationship between hosts and parasites in natural ecosystems are discussed. Chapter 10 also illustrates how patterns of movement and use of habitat space affect rates of parasitism in populations of *O. nuttalli* and *P. leucopus* (see Jennison et al. 2006 for details). Chapter 11 is unique in that it introduces a concept of landscape aesthetics in which aesthetics, as an emerging property of natural and cultural landscapes, influences resource recognition, management, conservation, and preservation of species such as the golden mouse. This chapter articulates the process in which the golden mouse has contributed to art, literature, and repositories of American culture and natural history (i.e., curio, museum). Finally, Chapter 12 outlines challenges and research opportunities as related to golden mouse landscape management, population genetics, and intraspecific and interspecific social relationships.

The Golden Mouse and the Levels-of-Organization Concept

Considering the levels-of-organization approach, the golden mouse (or your favorite small mammal species) can be effective in influencing educators, resource managers, and policy makers regarding the management of rare and little understood native species and their role in ecosystem dynamics. One has only to browse in a Barnes and Noble, Borders, or Waldens bookstore or to visit a Bass Pro Shop to realize that greater emphasis is placed on large mammals, such as polar bears (*Ursus maritimus*), timber wolves (*Canis lupus*), white-tailed deer (*Odocoileus virginianus*), and mountain lions (*Puma concolor*) — often referred to as "charismatic megafauna"—than on small mammals. Even intermediate-sized mesocarnivores or mesoomnivores such as raccoons (*Procyon lotor*), red foxes (*Vulpes vulpes*), bobcats (*Lynx rufus*), and striped skunks (*Mephitis mephitis*) command their share of shelf space, or DVDs. It is rare when a small mammal, such as the golden mouse, shares shelf space or documentary highlights in the bookstore. We hope this volume might help "level the playing field." The chapters that follow in this book will amplify this level-of-organization concept.

Organism Level

At the organismal level of organization one has only to study Figure 1.5 to appreciate the beauty and alert behavior of this unique, nocturnal species. It is not unusual to observe a golden mouse after it leaves its nest, then pausing on a nearby limb to groom. It will sit quietly on such a limb for a long period of time if not disturbed. It is also a time when a mammalogist or ecologist truly appreciates the beauty (rich golden color that is unique among cricetids) and, especially, the docile behavior of golden mice. Linzey and Packard (1977) noted that the golden mouse is unique in its burnished to golden color within the neotomyine–peromyscine group.

Golden mice prefer to live in a variety of habitats, but most frequently they occur in association with heavy undergrowth dominated by greenbrier, Chinese privet

FIGURE 1.5. An adult golden mouse (*O. nuttalli*). It is easy to see why early naturalists considered them the "most beautiful mouse." Notice how the prehensile tail aided the movement of the golden mouse as it moved down a stem of Chinese privet. Photograph by Thomas Luhring.

(*L. sinense*), Amur and Japanese honeysuckle (*Lonicera maackii* and *L. japonica*, respectively), and the Eastern red cedar. During 40 years of observing golden mice, my rule of thumb is "the heavier the undergrowth and thicket-like vegetation, the more likely one is to locate an active nest of golden mice"—such as the nest located in honeysuckle vines collected during a 2005 trip to Big Hill, Kentucky (Figure 1.6).

Population Level

A population is defined as any group of potentially interbreeding organisms occupying a particular space and time, and functioning as part of a biotic community. Population dynamics deals with factors and mechanisms that cause changes in

Figure 1.6. The globular nest of a golden mouse, built in honeysuckle vines, that was collected in Madison County, Kentucky. Photograph by Luis Rodas.

abundance and structure of populations in time and space. Those factors and mechanisms that pertain to golden mice are presented in detail in Chapter 3. Here I only intend to provide a couple of observations to illustrate the population level of ecological organization.

As noted, I conducted my first research dealing with small mammals in the Midwest (Barrett and Darnell 1967), where increased rates of reproduction were observed in the spring, followed by increased numbers during summer. Interestingly, in the southeastern United States, increased rates of reproduction occur during mid- to late winter, peak in early spring (typically early April for *O. nuttalli* and *P. leuco-pus* at the HSB Experimental Site) and then decrease through the summer and autumn (Christopher and Barrett 2006, Jennison et al. 2006). I suggest, based on lim- ited field observations, this decrease in abundance of golden mice is due to snake predation; avian predation appears to be minimal because of thick vegetative cover. For example, on numerous occasions, I have observed the Eastern rat snake (*Pantherophis alleghaniensis*) exploring golden mouse habitat where their nests were abundant. This verification, however, awaits further observation and research.

A second interesting observation at the population level is that golden mice appear to have specific habitat preferences (Goodpaster and Hoffmeister 1954, Linzey 1968), yet large areas of presumably suitable habitat might contain no individuals. McCarley (1958) suggested that the main factor controlling the abundance and distribution of this species was the density of underbrush. Although *O. nuttalli* are frequently found in heavy underbrush habitats, I have searched similar habitats for golden mice nests or activity to no avail. This observation represents another question requiring additional field research necessary to explain the population dynamics or rarity (Chapter 7) of this unique species.

Community Level

A community includes all of the populations inhabiting a specific area at the same time. Chapters 4 and 10 describe the golden mouse and its small mammal community relationships in greater detail, including the niche concept, and species interactions such as parasitism, competition, and coexistence. A couple of examples illustrate a community-level hierarchical approach.

In late November 1980, Barbara Knuth, then a graduate student at Miami University of Ohio, came upon a globular nest while searching for golden mice in Madison County, Kentucky. To our surprise, both a golden mouse and white-footed mouse occupied this nest. Because numerous mammalogists at that time argued that interspecies competition was the interaction that structured animal communities (e.g., Brown et al. 1979), I was not only excited but made sure to verify this rare observation. By the way, golden mice collected on this expedition became part of a resource partitioning study later published in the *Journal of Mammalogy* (Knuth and Barrett 1984). Barbara is currently Senior Associate Dean of the College of Agriculture and Life Sciences at Cornell University. Is there a relationship between earlier studies on golden mice and academic success and leadership?

This social interspecific interaction between *O. nuttalli* and *P. leucopus* is no longer a surprise. For example, we recently reported simultaneous double captures of these two species during the 2001 and 2004 field seasons at HSB (Christopher and Barrett 2007). The interspecific sociality of golden mice and white-footed mice provides a unique opportunity for field research in community ecology.

Another community-level observation involving golden mice is the composition of sympatric species of small mammals. During the past decade of extensive trapping at the HSB Experimental Site, the ranking in abundance of small mammals is consistently *P. leucopus* (first), *O. nuttalli* (second), *Glaucomys volans* (third), and *Tamias striatus* (fourth) (Christopher and Barrett 2006, Jennison et al. 2006). Why does apportionment remain constant for small mammals over an extended period of time in this riparian forest habitat in the southeastern United States? Is apportionment related to abundance or body mass, or are flying squirrels and chipmunks less likely to enter Sherman traps? More research is needed.

Ecosystem Level

An ecosystem is a biotic community and its abiotic environment functioning as a system, first defined by Tansley (1935). Chapter 5 focuses on the role of golden mice in ecological systems. Again, I will use only a couple of observations to illustrate the functioning of golden mice at the ecosystem level. I earlier discussed the importance of undergrowth (honeysuckle, privet, greenbrier) regarding its relationship to *O. nuttalli* distribution, abundance, and use of habitat space. The structure of an ecosystem is critical to the reproductive success and survivorship of this species. Chinese privet (*L. sinense*), a controversial invasive plant species, is important regarding habitat quality (fruits, cover) for golden mice. *L. sinense* provides extensive edge habitat along roads or fencerows frequently inhabited by *O. nuttalli* and is a major reason why golden mice could be considered an edge species where privet occurs.

Early reports in the literature note that sumac (*Rhus*) was found on golden mice "feeding platforms" (de Rogeot 1964, Goodpaster and Hoffmeister 1954). Indeed, these investigators reported that the most preferred seeds in the diet of golden mice were sumac. In a later feeding study, however, Jewell et al. (1991) reported that *O. nuttalli*, when fed seeds of smooth sumac (*Rhus glabra*), exhibited low rates of ingestion, excessive weight loss, and death of nine individuals during an 8-day feeding period. Feeding studies currently being conducted at HSB have confirmed that staghorn sumac (*R. typhina*) is seldom consumed when other dietary choices are made available. These findings also bring into question whether early observations (Goodpaster and Hoffmeister 1954) were actually "feeding platforms" of golden mice.

It is interesting that golden mice ingested and assimilated significantly more Japanese honeysuckle in their diet—an invasive exotic species—compared to a diet of the Eastern red cedar tree—a native species. Individuals on the juniper diet lost significantly more body mass and had lower survivorship than those on the honeysuckle diet. This difference was attributed to a higher mean crude protein in honeysuckle compared to red cedar (Peles et al. 1995). This food quality and dietary diversity relationship helps to explain the feeding behavior of the golden mouse in forest and disturbed ecosystems.

I question whether "feeding platforms" are ever used by *O. nuttalli*. In fact, during four decades of field observations in Georgia, South Carolina, and Kentucky, I have never found what I would term to be a golden mouse-feeding platform. Packard and Garner (1964), and Blus (1966) also failed to find feeding platforms. Golden mice likely were not consuming the abundance of sumac seeds on these "feeding platforms" as reported earlier. In summary, the structure of forest ecosystems, including vegetative cover, food diversity and quality, and quality of invasive plant species, plays a significant role in small mammal bioenergetics and reproductive success.

Landscape Level

A landscape is a heterogeneous area composed of a cluster of interacting ecosystems that are repeated in a similar manner throughout the region. Landscape ecology considers the development and dynamics of spatial heterogeneity, spatial and temporal interactions and exchanges across heterogeneous landscapes, influences

Golden mouse mother nursing her young that are only 3-4 days old. Photograph by Luis Rodas.

A golden mouse calmly grooming upon exiting a nest. This species is considered one of the most docile of all small mammal species. Photograph by Thomas Luhring.

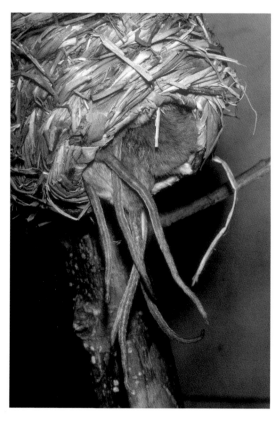

A "nest ball" of golden mice at The Swamp, Brookfield Zoological Park near Chicago, Illinois. Note the number of golden mouse tails extending from the communal ball of small mammal body mass. Photograph by Jim Schulz.

A white-footed mouse (*Peromyscus leucopus*) scurrying for cover in a riparian forest habitat. Golden mice and white-footed mice exhibit much niche overlap regarding their coexistence in this ecosystem-type habitat. Photograph by Thomas Luhring.

Globular nest of the golden mouse embedded on thick vegetative undergrowth. Photograph by Thomas Luhring.

Communal-shelter golden mouse nest situated in a honeylocust (*Gleditsia triacanthos*) tree, HorseShoe Bend Experimental Site, University of Georgia, Athens, Georgia. Photograph by Terry L. Barrett.

FIGURE 11.2. Diptych, *Portrait of Federico da Montefeltro and his spouse Battista Sforza*, circa 1465, by Piero della Francesca (c. 1420–1492) exemplary of figure–ground relationship. Image from Alinari/Art Resource, New York. Uffizi, Florence, Italy.

FIGURE 11.3A. Golden mouse with ear tag, 2005, HorseShoe Bend Experimental Site, University of Georgia, Athens, Georgia. Photograph by Thomas Luhring.

FIGURE 11.4B. Poppy, the deer mouse, donning earring from Ragweed, the golden mouse. Illustration by Brian Floca. Reprinted from AVI 2007a.

Plate XL

Dr. fron Nature by J.J. Audubon. P.R.S. F.L.S.

White Footed Mouse.

Lith. Printed & Col.d by J. T. Bowen. Philada.

FIGURE 11.5. Lithography entitled "White Footed Mouse," 1846, by John James Audubon, from the work of John James Audubon. Reproduced from Audubon 1989 with permission from Borders, Inc.

C

MACROCOSM

B

MESOCOSM

A

MICROCOSM

FIGURE 11.13. Increased scale of ecosystems. (A) Terrarium, HSB Ecological Research Center, University of Georgia, Athens, Georgia (microcosm). Photograph by Louis Rodas. (B) Breeding area, The Swamp, Brookfield Zoological Park, Chicago, Illinois (meso-cosm). Photograph by Jim Schulz, courtesy of the Chicago Zoological Society (C) HSB Experimental Site, riparian peninsula, North Oconee River, Athens, Georgia (macrocosm). Photograph, courtesy of the Eugene P. Odum School of Ecology, University of Georgia.

of spatial heterogeneity on biotic and abiotic processes, and management of spatial heterogeneity for society's benefit and survival (Risser et al. 1984). The landscape mosaic is composed of three major elements: matrix, patches, and corridors. Currently, there exists a paucity of information regarding how golden mice function at the landscape scale (i.e., golden mouse interactions at this scale provide an array of excellent research opportunities). Chapter 6 attempts to address what we do know about the social organization, dispersal, and colonization of golden mice in these fragmented landscapes.

There is some information, however, regarding the niche of golden mice at this scale. For example, golden mice seem to prefer not only dense vegetative undergrowth as described earlier, but also edge species such as Chinese privet and both Amur and Japanese honeysuckle. An edge is where two or more structurally different ecosystems meet; an edge species, such as *O. nuttalli*, inhabits edge or boundary habitats to meet reproductive or survivorship needs. We have located numerous *O. nuttalli* nests in this edge habitat over the past decade. Interestingly, the trunks of older mature (> 25 cm diameter breast height) privet are typically hollow, and both *O. nuttalli* and *P. leucopus* nest in these hollow cavities (Figure 1.7).

Golden mice are also frequently abundant in landscape patches dominated by the Eastern red cedar tree and thickets of Japanese honeysuckle (Peles et al. 1995). Landscape corridors, except in riparian habitats, seldom connect these patches. The riparian habitat created by the meander of a stream illustrates a natural resource corridor (Odum and Barrett 2005). The meander of the North

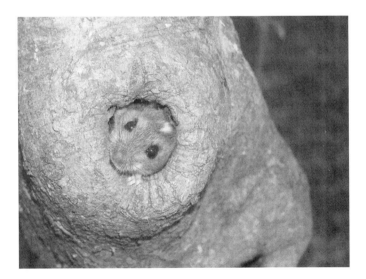

FIGURE 1.7. A white-footed mouse (*P. leucopus*) peering out of a nest cavity located in a large Chinese privet (*L. sinense*) bush—habitat that is also favored by golden mice. Photograph by Thomas Luhring.

FIGURE 1.8. Photograph showing how a meander in the North Oconee River creates a 35-acre (14.2-ha) experimental landscape peninsula. This peninsula is the site of the HorseShoe Bend experimental station used to investigate the population dynamics of small mammals, including *O. nuttalli*. Photograph courtesy of the Eugene P. Odum School of Ecology, University of Georgia.

Oconee River in North Georgia (Figure 1.8) creates a landscape peninsula with riparian habitat abundant on both sides of the stream. Although we know that individual white-footed mice were able to swim across this fifth-order stream and return to their original site of capture (Klee et al. 2004), we do not know if *O. nuttalli* of similar body mass and behavior exhibit this homing instinct. Neither do we know if individual golden mice use the riparian habitat as natural resource corridors. These landscape-scale questions, and many more, await further investigation.

Concluding remarks

We elected to use the level-of-organization hierarchy as a model approach in organizing the chapters in this book. We encourage readers to consider how biological, ecological, and evolutionary processes, including behavior, development, diversity, energetics, evolution, integration, and regulation, transcend these levels of organization.

Barrett and Peles (1999a) and Halle and Stenseth (2000) present reasons why small mammals—specifically rodents—are model organisms to address questions across levels of organization. Rodents are ideal because we often have detailed information regarding biology and natural history, we have the ability to monitor patterns of movement, and small mammals live in relatively small spatial areas, have short lives, high reproductive rates, rapid population turnover times, typically disperse from their natal areas upon reaching adulthood, and frequently exhibit response to seasonal change. The golden mouse fits most of these criteria as a model small mammal species when addressing questions across levels of organization. There are, however, a plethora of unusual characteristics exhibited by golden mice—nesting behavior, bioenergetics, patterns of movement—and interesting questions that require additional investigations at each level of organization, including sociality, rarity, and population regulation.

It is imperative that we better understand how species, such as the golden mouse, exist in a continuing fragmented landscape. For example, Hilty et al. (2006) discussed in their book entitled *Corridor Ecology* how the science and practices of linking patches and ecosystems conserve biotic diversity. Barrett and Peles (1999b) in their book entitled *Landscape Ecology of Small Mammals* also note the importance of investigating small mammals at the organismal, population, community, ecosytem, and landscape levels. In the chapters that follow, it is our intent to illustrate how golden mice have evolved numerous strategies, mechanisms, and behaviors that permit them to survive in both natural and disturbed ecosystems and landscapes—evolutionary relationships that perhaps human societies need to better understand on our ever-changing planet.

Literature Cited

Barbour, R.W. 1942. Nests and habitat of the golden mouse in Kentucky. Journal of Mammalogy 23:90–91.

Barbour, R.W., and W.H. Davis. 1974. Mammals of Kentucky. University Press of Kentucky, Lexington, Kentucky.

Barrett, G.W. 1968. The effects of an acute insecticide stress on a semi-enclosed grassland ecosystem. Ecology 49:1018–1035.

Barrett, G.W., and R.M. Darnell. 1967. Effects of dimethoate on small mammal populations. American Midland Naturalist 77:164–175.

Barrett, G.W., and J.D. Peles. 1999a. Small mammal ecology: A landscape perspective. Pages 1–10 *in* G.W. Barrett and J.D. Peles, editors. Landscape ecology of small mammals. Springer-Verlag, New York, New York.

Barrett, G.W., and J.D. Peles, editors. 1999b. Landscape ecology of small mammals. Springer-Verlag, New York, New York.

Barrett, G.W., J.D. Peles, and E.P. Odum. 1997. Transcending processes and the levels-of-organization concept. BioScience 47:531–535.

Blesh, J., and M. Williams. 2003. HorseShoe Bend: A center for ecological teaching, research, and service. Institute of Ecology, University of Georgia, Athens, Georgia.

Blus, L.J. 1966. Some aspects of golden mouse ecology in southern Illinois. Transactions of the Illinois State Academy of Science 59:334–341.

Brown, J.C., O.J. Reichman, and D.W. Davidson. 1979. Granivory in desert rodents. Annual Review of Ecology and Systematics 10:210–227.

Christopher, C.C., and G.W. Barrett. 2006. Coexistence of white-footed mice (*Peromyscus leucopus*) and golden mice (*Ochrotomys nuttalli*) in a southeastern forest. Journal of Mammalogy 87:102–107.

Christopher, C.C., and G.W. Barrett. 2007. Double captures of *Peromyscus leucopus* and *Ochrotomys nuttalli*. Southeastern Naturalist. In press.

de Rogeot, R.M. 1964. The golden mouse. Virginia Wildlife 25:10–11.

Dietz, B.A., and G.W. Barrett. 1992. Nesting behavior of *Ochrotomys nuttalli* under experimental conditions. Journal of Mammalogy 73:577–581.

Feldhamer, G.A., and K.A. Maycroft. 1992. Unequal capture responses of sympatric golden mice and white-footed mice. American Midland Naturalist 128:407–410.

Goodpaster, W.W., and D.F. Hoffmeister. 1954. Life history of the golden mouse, *Peromyscus nuttalli*, in Kentucky. Journal of Mammalogy 35:16–27.

Halle, S., and N.C. Stenseth. 2000. Introduction. Pages 3–17 *in* S. Halle and N.C. Stenseth, editors. Activity patterns in small mammals: An ecological approach. Springer-Verlag, Berlin, Germany.

Hendrix, P.F. 1997. Long-term patterns of plant production and soil carbon dynamics in a Georgia piedmont agroecosystem. Pages 235–245 *in* E.A. Paul, K. Paustian, E.T. Elliott, and C.V. Cole, editors. Soil organic matter in temperate agroecosystems. CRC Press, Boca Raton, Florida.

Hilty, J.A., W.Z. Lidicker, Jr., and A.M. Merenlender. 2006. Corridor ecology: The science and practice of linking landscapes for biodiversity conservation. Island Press, Washington, DC.

Jennison, C.A., L.R. Rodas, and G.W. Barrett. 2006. *Cuterebra fontinella* parasitism on *Peromyscus leucopus* and *Ochrotomys nuttalli*. Southeasten Naturalist 5:157–164.

Jewell, M.A., M.K. Anderson, and G.W. Barrett. 1991. Bioenergetics of the golden mouse on experimental sumac seed diets. American Midland Naturalist 125:360–364.

Klee, R.V., A.C. Mahoney, C.C. Christopher, and G.W. Barrett. 2004. Riverine peninsulas: An experimental approach to homing in white-footed mice (*Peromyscus leucopus*). American Midland Naturalist 151:408–413.

Knuth, B.A., and G.W. Barrett. 1984. A comparative study of resource partitioning between *Ochrotomys nuttalli* and *Peromyscus leucopus*. Journal of Mammalogy 65:576–583.

Linzey, D.W. 1968. An ecological study of the golden mouse, *Ochrotomys nuttalli*, in the Great Smoky Mountains National Park. American Midland Naturalist 79:320–345.

Linzey, D.W., and R. L. Packard. 1977. *Ochrotomys nuttalli*. Mammalian Species 75:1–6.

McCarley, W.H. 1958. Ecology, behavior and population dynamics of *Peromyscus nuttalli* in eastern Texas. Texas Journal of Science 10:147–171.

Odum, E.P., and G.W. Barrett. 2005. Fundamentals of ecology, 5th ed. Thomson Brooks/Cole, Belmont, California.

Packard, R.L., and H. Garner. 1964. Aboreal nests of the golden mouse in eastern Texas. Journal of Mammalogy 45:369–374.

Peles, J.D., C.K. Williams, and G.W. Barrett. 1995. Bioenergetics of golden mice: The importance of food quality. American Midland Naturalist 133:373–376.

Pruett, A.L., C.C. Christopher, and G.W. Barrett. 2002. Effects of a forested riparian peninsula on mean home range size of the golden mouse (*Ochrotomys nuttalli*) and the white-footed mouse (*Peromyscus leucopus*). Georgia Journal of Science 60:201–208.

Risser, P.G., J.R. Karr, and R.T.T. Forman 1984. Landscape ecology: Directions and approaches. Natural History Survey Special Publication 2, Champaign, Illinois.

Springer, S.D., P.A. Gregory, and G.W. Barrett. 1981. Importance of social grouping on bioenergetics of the golden mouse, *Ochrotomys nuttalli*. Journal of Mammalogy 62:628–630.

Stueck, K.L., M.P. Farrell, and G.W. Barrett. 1977. Ecological energetics of the golden mouse based on three laboratory diets. Acta Theriologica 22:309–315.

Tansley, A.G. 1935. The use and abuse of vegetational concepts and terms. Ecology 16:284–307.

Terres, J.K. 1966. Search for the golden mouse. Audubon Magazine 68:96–101.

2
The Golden Mouse: Taxonomy and Natural History

GEORGE A. FELDHAMER AND DONALD W. LINZEY

In symmetry of form and brightness of colour, this is the prettiest species of Mus inhabiting our country. (Audubon and Bachman 1841:99)

Although the designation *Mus* has long been abandoned as the generic name for the golden mouse (*Ochrotomys nuttalli*), the sentiments of Audubon and Bachman (1841) capture the appeal of this species for biologists today, and its morphology, life history, and ecological adaptations continue to be active subjects of research. In many respects, the golden mouse represents a striking contrast to more intensively studied species such as the white-footed mouse (*Peromyscus leucopus*) and the deer mouse (*P. maniculatus*). Whereas these species of *Peromyscus* can be found in nearly every terrestrial habitat in North America, the golden mouse is much less widespread geographically, generally is more habitat-specific, is less abundant locally, and certainly has not received as much attention from investigators. In these respects, it is representative of the vast majority of mammalian species in North America (Feldhamer and Morzillo Chapter 7 of this volume, Gaston 1994).

In this chapter, we discuss basic aspects of the taxonomy, life history, and ecology of the golden mouse to set the stage for the more detailed chapters to follow that focus on the golden mouse at different levels of ecological organization.

Geographic Distribution

The golden mouse is restricted to the southeastern United States from the Appalachian Mountains in northwestern Virginia south to central Florida and from extreme eastern Texas and Oklahoma through southeastern Missouri to southern Illinois and most of Kentucky (Figure 2.1). It has been recorded from a seepage bog on Andrews Bald in the Great Smoky Mountains National Park, Swain County, North Carolina, at an elevation of 5800 ft (1761 m), the highest known elevation for this species (Linzey et al. 2002).

Although often found in deciduous hardwood and coniferous forests, the golden mouse occupies a variety of habitats, including the borders of old fields, swampy

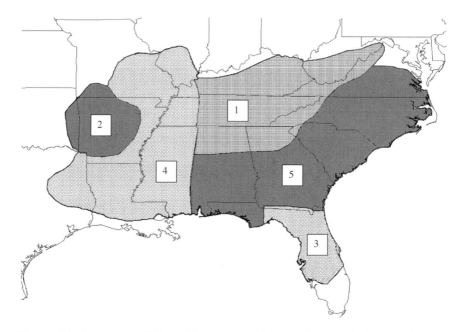

FIGURE 2.1. Range map of the golden mouse with currently recognized subspecies: (1) *O. n. aureolus*; (2) *O. n. flammeus*; (3) *O. n. floridanus*; (4) *O. n. lisae*; and (5) *O. n. nuttalli.* Adapted from Packard (1969).

lowlands, canebrakes, and xeric wooded uplands. Its occurrence is usually associated with the presence of abundant climbing vines, such as greenbrier (*Smilax* sp.), Japanese honeysuckle (*Lonicera japonica*), grape (*Vitis* sp.), and poison ivy (*Toxicodendron radicans*) and dense understory vegetation, such as cane (*Arundinaria* sp.) and blackberry (*Rubus* sp.). It is regarded as a habitat specialist (Dueser and Hallett 1980, Knuth and Barrett 1984, Seagle 1985). Although it might be fairly common in localized areas, as a general rule the golden mouse is uncommon with population densities well below those of sympatric species of *Peromyscus* (Feldhamer and Morzillo Chapter 7 of this volume), even after consideration of its reduced trappability caused by arboreal habits.

Classification

Generic and Species-level Taxonomy: The Early Years

Richard Harlan, a physician and naturalist, first introduced the golden mouse to the scientific community in 1832 based on a specimen taken near Norfolk, Norfolk County, Virginia. He placed it in the genus *Arvicola*, with the specific epithet *nuttalli* in honor of Thomas Nuttall, "the eminent botanist" who had collected the specimen. Unfortunately, Harlan's description was fairly general and he failed to note any of the truly diagnostic morphological characteristics of the golden mouse. Nine years later, John J. Audubon and Rev. John Bachman (1841) named an "orange-coloured mouse" taken from an oak forest in South Carolina as *Mus aureolus* ("golden") and placed it in the subgenus *Calomys* ("beautiful mouse"). Spencer F. Baird (1857), at the time the Assistant Secretary of the Smithsonian Institution, considered specimens of the "red mouse" from throughout much of the range as *Hesperomys* ("western mouse") *nuttalli*. He noted (1857:468) that it was with ". . . very decided regret that I am impelled, by a strict regard for the law of priority, to change the expressive name of *aureolus* . . . for the less meaning one of *nuttalli*. There can, however, be little, if any, doubt that the species described by Harlan, in 1832, belongs to the present animal." As late as 1885, however, Frederick True (1885), Curator of the Department of Mammals at the Smithsonian Institution, referred to this species as *Hesperomys aureolus* (Audubon and Bachman 1841) Wagner. Bangs (1898) placed the golden mouse in the genus *Peromyscus*, which was retained by Osgood (1909) as *P. nuttalli* in his monumental revision of the genus. However, as previous investigators had done before him, Osgood described several unique characteristics that set the golden mouse apart from other species of *Peromyscus*. In addition to the fact that, unlike other *Peromyscus*, the ears of the golden mouse are the same golden color as the dorsal pelage and juveniles have a similar pelage color as the adults, Osgood (1909:223) noted: "It is rather surprising that the numerous characters of *P. nuttalli* have not been accorded more than specific rank. It differs widely from all other species of the genus in external, cranial, and dental characters." Although he retained the golden mouse in the genus *Peromyscus*, Osgood considered it the sole member of the subgenus *Ochrotomys* ("yellow mouse"), one of six subgenera he designated. He stated (1909:26) that "*Ochrotomys* is a subgenus based upon a single aberrant form (*P. nuttalli*), which seems to have no very close relative, although the general characters are obviously those of *Peromyscus*."

Several other investigators also suggested that features of the golden mouse were different enough to warrant a separate generic status, including Blair (1942), who was the first to use characteristics of the baculum (*os penis*) to determine relationships among species of *Peromyscus*. He found that the baculum of the golden mouse was significantly smaller and shorter than any other species in the genus. On the basis of this and other morphological differences, Blair (1942:201) concluded that "*Ochrotomys* is so different from the other *Peromyscus* as to raise the question as to whether its relationships might be better shown by elevating *Ochrotomys* to generic rank." However, several later investigators, including Hall and Kelson (1959) in their seminal work on North American mammals, retained the golden mouse within the genus *Peromyscus*. Nonetheless, various lines of evidence continued to accumulate suggesting that elevation of the golden mouse to generic status was warranted.

Hooper (1958:23) examined the male genitalia, specifically the glans penis, in *Peromyscus* and found that the urn-shape and large spines of the golden mouse phallus were distinctive. Along with other factors, this convinced him that the species "should be raised to generic level."

Hirth (1960) noted that golden mouse sperm was distinctly different from that of five other *Peromyscus* species studied. Subsequent work on male genitalia by Hooper and Musser (1964) reinforced this conclusion. Manville (1961:104) stated that *Ochrotomys* "deserves recognition as generically distinct" based on the structure of the entepicondylar foramen. This small opening on the margin of the distal end of the humerus occurs in *Peromyscus* (Rinker 1960) but not in the golden mouse. Studies of the male reproductive tract by Hooper and Musser (1964) and Arata (1964) also supported the conclusion that the golden mouse warranted separate generic status.

Ontogenetic characteristics also serve to differentiate the golden mouse. Although based on a limited number of specimens, Layne (1960) found that the growth and development of golden mice was much more rapid than comparably sized species of *Peromyscus*. The eruption of the incisors, the opening of the eyes, and the erection of the pinnae occur at an earlier age in *Ochrotomys* than in *Peromyscus*. Linzey and Linzey (1967b) substantiated these results in their study of growth and development of 108 litters. Layne (1960) concluded that its relatively accelerated ontogeny and certain behavioral patterns of the young, such as a tendency to cling tenaciously at an early age, to be less mobile, and to freeze when disturbed were adaptations for arboreal nesting. The karyotype of the golden mouse is also distinct from that of *Peromyscus*. Based on four specimens, Patton and Hsu (1967) found that the golden mouse has a diploid number of 52 compared with 48 in *Peromyscus* and concluded that the chromosomal data, in conjunction with other criteria, confirmed earlier views that *Ochrotomys* merited generic status. Presently, the golden mouse is considered the sole member of the genus *Ochrotomys*.

On the basis of the morphology of the male reproductive system of muroid rodents, Arata (1964) recognized four groups of genera with simple phallic form, of which *Ochrotomys, Baiomys* (pygmy mice), and *Onychomys* (grasshopper mice) comprised one group and *Peromyscus* and *Reithrodontomys* (harvest mice) comprised another. Hooper (1968) grouped *Ochrotomys* with the genera *Peromyscus, Reithrodontomys, Onychomys, Baiomys, Scotinomys* (brown mice), and *Neotomodon alstoni* (the Mexican volcano mouse) as a distinct phyletic line of New World rodents—the "peromyscines." Packard (1969) suggested that the golden mouse was a primitive member of this group. Likewise, based on morphological analyses, Carleton (1980) considered the golden mouse as the basal clade to peromyscines, whereas Patton et al. (1981) reached a similar conclusion from allozyme data. Engstrom and Bickham (1982:123), however, in a study of G-bands of chromosomes found " . . . a general lack of chromosomal homology between golden mice and other peromyscine genera." In fact, there was such a divergence in the karyotype of *O. nuttalli* from that of *Peromyscus* that they suggested the golden mouse had undergone "karyotypic megaevolution" (*sensu* Baker and Bickham 1980) and was more closely related to *Sigmodon* (the cotton rats) within a clade of "neotomines."

Family-Level Changes and Current Taxonomy

The current taxonomy of the golden mouse reflects recent major changes and reevaluations in the long-standing debate on the phylogeny of rodents (Carleton and Musser 2005). Our current view of rodent taxonomy has changed considerably, due in large part to genetic sequence data. Nonetheless, questions and unresolved issues concerning the validity and relationships of various families of rodents still remain. Musser and Carleton (1993) included the golden mouse in the family Muridae. At that time, with about 280 genera and 1300 species worldwide, murids not only comprised the largest family of mammals, but they also encompassed about 28 percent of all extant mammalian species. Given such a huge assemblage of species, it is not surprising that 17 subfamilies of murids were proposed. Along with other New World rats and mice, the golden mouse was placed in the subfamily Sigmodontinae. More recently, Musser and Carleton (2005), in their reevaluation of rodent classification based on morphological characteristics and molecular analyses (Jansa and Weksler 2004, Marten et al. 2000, Michaux et al. 2001), reduced the murids to 150 genera with 730 species. They reinstated five families that were earlier included in the Muridae: the Platacanthomyidae (tree mice), Spalacidae (zokors, bamboo rats, mole rats), Calomyscidae (mouselike hamsters), Nesomyidae (pouched rats and mice), and Cricetidae (rats and mice). These five families and the Muridae comprised the Superfamily Muroidea. Musser and Carleton (2005:896) noted that they "steered a conservative course in recognizing just six families" in this group. The family Cricetidae includes six subfamilies (Table 2.1), one of which, the Neotominae, includes among its 14 genera both the monotypic *Ochrotomys* and the speciose *Peromyscus* plus 12 other genera. The complete classification of *Ochrotomys* based on Musser and Carleton (2005) is shown in Table 2.1.

Subspecies Designations

As greater numbers of specimens of the golden mouse were collected, the number of presumptive "subspecies" increased. Howell (1939) described *P. n. lewisi* from Amelia County, Virginia. It was named for J. B. Lewis, one of Virginia's outstanding scientists. Two years later, Goldman (1941) recognized another "subspecies," *P. n. flammeus*, from Pike County, Arkansas. Along with the subspecies *nuttalli* from the original description of the golden mouse (Harlan 1832) and *aureolus* of Audubon and Bachman (1841), these were the four "subspecies" recognized by Hall and Kelson (1959). Packard (1969) recognized five "subspecies" of the golden mouse based on a comprehensive analysis of external and cranial measurements and color variation. He retained *flammeus*, *nuttalli*, and *aureolus* as "subspecies" but considered *lewisi* as synonymous with *nuttalli*. He also described two new "subspecies": *O. n. floridanus* (type locality Welaka, Putnam County, Florida) and *O. n. lisae* (type locality Nacogdoches, Nacogdoches County, Texas). The geographic ranges of the presumptive "subspecies" recognized by Packard (1969) differed (Figure 2.1) from those of Hall and Kelson (1959) but were then adopted by Hall (1981) in his revision.

TABLE 2.1. Classification of the rodents and placement of the golden mouse (boldface taxa).

Order **Rodentia**
Suborder **Myomorpha**
Anomaluromorpha
Castorimorpha
Hystricomorpha
Sciuromorpha
Superfamily **Muroidea**
Dipodoidea
Family **Cricetidae**
Calomyscidae
Muridae
Nesomyidae
Platacanthomyidae
Spalacidae
Subfamily **Neotominae**
Arvicolinae
Cricetinae
Lophiomyinae
Sigmodontinae
Tylomyinae
Tribe **Ochrotomyini**
Baiomyini
Neotomini
Reithrodontomyini
Genus ***Ochrotomys***
Species ***nuttalli***

Source: After Musser and Carleton (2005).

If a subspecies is viewed as an evolutionary unit in which gene flow is reduced, then the five "subspecies" of *Ochrotomys* should be viewed as geographic variants—not subspecies—because they intergrade one into another. The biological subspecies concept (*sensu* Whitaker 1970) recognizes reduced gene flow as the primary criterion for determination of subspecies. Nonetheless, individuals from different groups potentially can interbreed because secondary isolating mechanisms have not formed completely (Whitaker 1970). Also, after a primary isolating mechanism is in effect, groups must have formed morphological or other differences. As noted by Whitaker and Hamilton (1998:320) for the golden mouse: "The described subspecies appear to intergrade one into another, and we therefore see no reason to distinguish them." We suggest that all golden mice should be classified as *Ochrotomys nuttalli nuttalli* (Harlan).

Morphology

As noted, Harlan's (1832) original description of the golden mouse included none of the unique characteristics that differentiate it from other rodent species, particularly members of the genus *Peromyscus*. One of the most obvious diagnostic characteristics of the golden mouse is the soft, thick pelage of "rich, burnished, ochraceous to golden color" (Whitaker and Hamilton 1998:317) on the dorsum. Slight regional differences might occur in the amount of reddish, yellowish, or brownish overtones of the dorsal pelage (Linzey and Packard 1977) of different geographic variants. The feet and underparts range from creamy white to cinnamon orange. Unlike *Peromyscus*, in which juvenile pelage is gray, young golden mice are essentially the same color as adults, only slightly duller. The tail is weakly bicolored and nearly equal in length to the head and body. It serves as a balancing organ when running along branches and is wrapped around a branch for additional support when the animal rests. The feet are smaller than

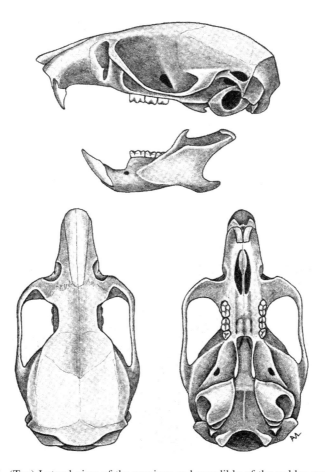

Figure 2.2. (Top) Lateral view of the cranium and mandible of the golden mouse; (bottom left) dorsal view of the cranium; (bottom right) ventral view of the cranium. From Linzey and Packard 1977; reproduced by permission of Alliance Communications Group, a division of Allen Press, Inc.

Peromyscus, with each fore and hind foot possessing six plantar tubercles. A rudimentary seventh tubercle is present at the base of the fifth digit. Foot structure might enhance the climbing ability of this highly arboreal species. There are a total of six mammae: two inguinal pairs and a pectoral pair.

Average measurements given by Whitaker and Hamilton (1998) are as follows: total length, 176 mm (range: 151–200); tail 85 mm (range: 51–97); hind foot, 20 mm (range: 16–21); weight 20–26 g. They noted, however, that in the southern parts of its range, *Ochrotomys* was smaller. Average measurements of 28 adults from Florida were as follows: total length, 158 mm (range: 135–178); tail, 72 mm (range: 60–85); hind foot, 18 mm (range: 13–19).

The skull of the golden mouse is relatively broad with an inflated braincase. It appears somewhat squarish in dorsal view (Figure 2.2) with a convex anterior edge of the zygomatic plate. Osgood (1909:223) noted that the "posterior palatine foramina are farther back than in *Peromyscus*, being decidedly nearer the interpterygoid fossa than to the posterior endings of the anterior palatine slits." The dental formula is typical of most rodents: 1/1, 0/0, 0/0, 3/3 = 16. The molars are short-crowned (brachyodont) with well-developed lophs and styles, and the enamel is thicker, with more compressed folds than in *Peromyscus*.

In addition to the many morphological features previously noted to differentiate *Ochrotomys* from *Peromyscus*, golden mice have a urethral process and prominent preputial glands, but they lack ampullary glands and a gallbladder (Arata 1964, Carleton 1980, Hooper 1958). More detailed diagnoses of anatomical, skeletal, cranial, and dental characteristics of the golden mouse are in Hall (1981), Linzey and Packard (1977), and Musser and Carleton (2005).

A Brief Overview of Ecology and Reproduction

Behavior

Golden mice are active during all seasons. They are nocturnal and spend the daylight hours in their nest. They are quite sociable, with as many as eight individuals sharing the same nest. They are adept at moving about among vines and small branches where they use their semi-prehensile tail for both balance and support. Whenever the animal pauses, its tail will immediately encircle a nearby vine or branch.

The adjective most commonly applied to this species is "docile." When handled, wild individuals will rarely bite or exhibit any aggressive action. Upon release, they often will scurry up the nearest tree to a nest or will sit quietly on a limb until the intruder leaves the area. Behavioral, neuromuscular, and vocal development of neonates is more rapid than in *Peromyscus*. Young can walk by 1 week of age and they show a tendency toward climbing soon afterward (Layne 1960, Linzey and Linzey 1967b). Space use seems restricted, but individuals move with amazing agility among thick vines and along branches, often high above the ground.

Populations

The golden mouse lives in highly localized populations. Population densities reported in the literature are highly variable (Linzey and Packard 1977). Reported densities have ranged up to nearly 7 per hectare (2.8 per acre) in Louisiana (Shadowen 1963) and nearly 9 per hectare (3.6 per acre) in Tennessee (Linzey 1968). Density is usually lower than that of syntopic *Peromyscus*, as might be expected when considering specialist versus generalist species. Also, there might be inverse relationships between population densities of golden mice and co-occurring *Peromyscus*. For example, Linzey (1968) found significant inverse relationships among *Ochrotomys*, *P. maniculatus*, and *P. leucopus* in Tennessee. Furtak-Maycroft (1991) found a similar inverse relationship between the number of golden mice and white-footed mice captured on short leaf pine (*Pinus echinata*) and loblolly pine stands (*P. taeda*) in southern Illinois. Pearson (1953) and McCarley (1958) reported reciprocal densities between golden mice and cotton mice (*P. gossypinus*). Conversely, absence of competitive interaction was suggested by the study of Christopher and Barrett (2006), in which relative abundance of golden mice did not change after removal of white-footed mice. Removal of white-footed mice from forested sites in southern Illinois also had little effect on space use of golden mice (Corgiat 1996). Feldhamer and Maycroft (1992) found that individual golden mice were trapped significantly fewer times, and in fewer traps, than syntopic white-footed mice. Also, both species were taken at the same station much less than expected, probably reflecting behavioral differences as well as potential interspecific competition. Home ranges are generally small, ranging from 0.1 to 1.5 acres (0.05 to 0.6 ha) (Blus 1966, Goodpaster and Hoffmeister 1954, Linzey 1968) but should certainly be considered three dimensional given the highly arboreal habits of golden mice. Individuals are sedentary (Komarek 1939) and can show a strong affinity for their home range despite major disturbances such as flooding (McCarley 1959). Golden mice inhabited forested areas less than a year after severe disturbance from cutting and burning (Furtak-Maycroft 1991). No doubt a mosaic of interacting biotic and abiotic factors affects spatial activity as well as potential competitive interactions in the golden mouse and suggests caution in interpreting single-factor analyses. Rose (Chapter 3 of this volume) expands on aspects of the population dynamics of the golden mouse, and Christopher and Cameron (Chapter 4 of this volume) more fully explore the question of potential interspecific competition in this species.

Feeding Habits

Golden mice are omnivorous and feed on a variety of seeds as well as invertebrates (Goodpaster and Hoffmeister 1954, Linzey 1968). Several common species of seeds recovered from remains in arboreal nests include wild cherry (*Prunus* sp.), dogwood (*Cornus* sp.), greenbrier (*Smilax* spp.), sumac (*Rhus* sp.), and oak (*Quercus* sp.). Seeds and other food are transported in internal cheek pouches.

Stomach analyses from Tennessee revealed invertebrate remains in up to 57 percent of the stomachs examined (Linzey 1968). Peles and Barrett (Chapter 8 of this volume) discuss various aspects of energy use and metabolism in the golden mouse.

Reproduction and Development

The breeding season varies somewhat depending on latitude, but it generally extends from early spring to autumn. In southern portions of its range, *Ochrotomys* might breed throughout the year but with reduced activity during the hottest part of the summer. As in *Peromyscus*, gestation is about 4 weeks with litter sizes of two to four. Significant differences were found between spring (2.4) and autumn (3.1) litter sizes in Tennessee (Linzey and Linzey 1967b). In addition, spring-born neonates were heavier and had smaller measurements than fall-born animals. As noted, young golden mice are better developed at birth than many of their counterparts. The prehensile nature of their tail is evident by the second day. The adult pelage is attained by a single maturational molt that begins in males at an average age of 36 days and in females at an average age of 38 days. The average duration of molt for both sexes is 29 days and 25 days, respectively (Linzey and Linzey 1967a).

Longevity

In captivity, Linzey and Linzey (1967b) recorded that 1 female produced 17 litters in 18 months. They also recorded females as old as 6.5 years of age bearing young. The average life span in the wild is probably less than a year, although life spans of at least 2.5 years have been reported (Bohall-Wood and Layne 1986, Linzey 1968, McCarley 1958, Pearson 1953). Linzey (1998) documented five captive golden mice that lived for 6 years or longer, with one female living 8 years, 5 months—the longest known life span ever recorded for a small rodent.

Predation

Potential predators include snakes, hawks, owls, and mammals such as the red fox (*Vulpes vulpes*), gray fox (*Urocyon cinereoargenteus*), mink (*Mustela vison*), skunks (*Mephitis mephitis* and *Spilogale putorius*), and long-tailed weasel (*Mustela frenata*). The semiarboreal habits of golden mice might make them less vulnerable to most of these predators than are the more terrestrial species of small mammals. The extent to which predation has an impact on populations of golden mice is unknown.

Parasites

In comparison to other cricetid rodents, relatively few parasites have been recorded from *O. nuttalli*. Internal parasites that have been identified include the following: Bacteria—*Grahamella* sp. and *Escherichia coli*; Cestoda—*Taenia rileyi*; and Nematoda—*Longistriata* sp. and *Rictularia* sp. (Linzey 1968, 1995).

Durden (Chapter 10 of this volume) discusses ectoparasites of the golden mouse and the role that the species plays in vector-borne diseases.

Nests

Arboreal Nests

The golden mouse is well known for building globular or somewhat elongated arboreal nests. Nests are especially easy to locate during winter after leaf fall, but they might be less conspicuous during other seasons. Various investigators have described and studied arboreal nests (Barbour 1942, Black 1936, Blus 1966, Frank and Layne 1992, Goodpaster and Hoffmeister 1954, Howell 1921, Layne 1958, Linzey 1968, Moore 1946). Nests from Tennessee and North Carolina averaged 5.4 in (13.7 cm) in length and 4.3 in (10.9 cm) in width. They weighed an average of 20.7 g. Nests might be located only a few centimeters off the ground or up to 30 ft (10 m) high in shrubs, trees, or vines. Most, however, are 5–15 ft (1.5–4.5 m) high. The average height for 32 nests in Tennessee and North Carolina was 6.4 ft (2.0 m), with a range from 1.5 to 27.0 ft (0.5–8.2 m). Nests in Tennessee and North Carolina were generally found in either pine (*Pinus* spp.) or Eastern red cedar trees (*Juniperus virginiana*). They were located in forks of the trees or in *Smilax* vines alongside the trunks. Two nests were in blackberry (*Rubus* spp.) patches (Linzey 1968). In Florida, arboreal nests are often suspended in clumps of Spanish moss (*Tillandsia usneoides*) and have also been found between fronds of saw palmetto (*Serenoa repens*) (Bohall-Wood and Layne 1986).

The nest, with a single small opening, usually consists of an outer layer of deciduous or coniferous leaves and an inner chamber lined with grasses and shredded bark. Birds' nests are sometimes remodeled. Several adult golden mice might occupy an arboreal nest at the same time. Communal nesting is especially evident during winter. Likewise, an individual might use more than one arboreal nest (Luhring and Barrett, Chapter 9 of this volume, Morzillo et al. 2003) in addition to using ground nests.

Ground Nests

In contrast to the extensive literature on arboreal nests of golden mice, there are few descriptions of ground nests used by this species. This certainly reflects the relative ease of locating arboreal nests compared to ground nests, which might be built in or under woody debris (McCarley 1958, Strecker and Williams 1929) or located underground. In some portions of the range, ground nests appear to be much more common than arboreal nests. The use of radiotelemetry and miniaturized transmitters greatly facilitates the study of ground nests. For example, Frank and Layne (1992) found 75 nests beneath deep litter and only 2 in shrubs aboveground. Aboveground and ground nests were similar in construction. Ground and arboreal nests used by an individual golden mouse might be quite far apart from each other (Morzillo et al. 2003).

In addition to arboreal nests, golden mice might also use elevated feeding platforms scattered throughout their home range. These platforms, described by Barbour (1942) and Goodpaster and Hoffmeister (1954), might be old bird nests or abandoned, degraded mouse nests. They are recognized by the litter of seed hulls from a variety of plants and other debris from past feeding. Just as individuals might share arboreal nests, several golden mice might use the same feeding platform.

Although the golden mouse appears to have no positive or negative economic significance, it is, nevertheless, a characteristic and attractive mammal of forest ecosystems of the southeastern United States. Now, 175 years after its discovery, there are still many aspects of its life history that remain to be discovered. In addition to the opportunities for further understanding of its own life history and ecology, this "beautiful mouse" provides an excellent model to explore broader questions of population dynamics, community interactions, and ecological relationships at all levels of organization.

Literature Cited

Arata, A.A. 1964. The anatomy and taxonomic significance of the male accessory reproductive glands of muroid rodents. Bulletin of the Florida State Museum, Biological Sciences 9:1–42.

Audubon, J.J., and J. Bachman 1841. Descriptions of new species of quadrupeds inhabiting North America. Proceedings of the Academy of Natural Sciences of Philadelphia 1:92–103.

Baird, S.F. 1857. General report on the mammals of the Pacific Railroad surveying parties. Part I in The mammals of North America: The descriptions of species based chiefly on the collections in the museum of the Smithsonian Institution, 3 parts. J.B. Lippincott and Co., Philadelphia, Pennsylvania.

Baker, R.J., and J.W. Bickham, 1980. Karyotypic evolution in bats: Evidence of extensive and conservative chromosomal evolution in closely related taxa. Systematic Zoology 29:239–253.

Bangs, O. 1898. The land mammals of peninsular Florida and the coastal region of Georgia. Proceedings of the Boston Society of Natural History 28:157–235.

Barbour, R.W. 1942. Nests and habitat of the golden mouse in eastern Kentucky. Journal of Mammalogy 23:90–91.

Black, J.D. 1936. Mammals of northwestern Arkansas. Journal of Mammalogy 17:29–35.

Blair, W.F. 1942. Systematic relationships of Peromyscus and several related genera as shown by the baculum. Journal of Mammalogy 23:196–204.

Blus, L.J. 1966. Some aspects of golden mouse ecology in southern Illinois. Transactions of the Illinois State Academy of Science 59:334–341.

Bohall-Wood, P., and J.N. Layne. 1986. The golden mouse. Florida Wildlife 49:16–19.

Carleton, M.D. 1980. Phylogenetic relationships in neotomine–peromyscine rodents (Muroidea) and a reappraisal of the dichotomy within New World cricetinae. Miscellaneous Publication 157, Pages 1–146. Museum of Zoology, University of Michigan, Ann Arbor, Michigan.

Carleton, M.D., and G.G. Musser. 2005. Order Rodentia. Pages 745–752 *in* D.E. Wilson and D.M. Reeder, editors. Mammal species of the world: A taxonomic and geographic reference, 3rd ed., 2 volumes. Johns Hopkins University Press, Baltimore, Maryland.

Christopher, C.C., and G.W. Barrett. 2006. Coexistence of white-footed mice (*Peromyscus leucopus*) and golden mice (*Ochrotomys nuttalli*) in a southeastern forest. Journal of Mammalogy 87:102–107.

Corgiat, D.A. 1996. Golden mouse microhabitat preference: Is there interspecific competition with white-footed mice in southern Illinois? MS thesis, Southern Illinois University, Carbondale, Illinois.

Dueser, R.D., and J.G. Hallett. 1980. Competition and habitat selection in a forest-floor small mammal fauna. Oikos 35:293–297.

Engstrom, M.D., and J.W. Bickham. 1982. Chromosome banding and phylogenetics of the golden mouse, *Ochrotomys nuttalli*. Genetica 59: 119–126.

Feldhamer, G.A., and K.A. Maycroft. 1992. Unequal capture response of sympatric golden mice and white-footed mice. American Midland Naturalist 128:407–410.

Frank, P.A., and J.N. Layne. 1992. Nests and daytime refugia of cotton mice (*Peromyscus gossypinus)* and golden mice (*Ochrotomys nuttalli)* in south-central Florida. American Midland Naturalist 127:21–30.

Furtak-Maycroft, K.A. 1991. An empirically-based habitat suitability index model for golden mice, *Ochrotomys nuttalli*, on pine stands in the Shawnee National Forest. MS thesis, Southern Illinois University, Carbondale, Illinois.

Gaston, K.J. 1994. Rarity. Chapman & Hall, New York, New York.

Goldman, E. A. 1941. A new western subspecies of golden mouse. Proceedings of the Biological Society of Washington 54:189–192.

Goodpaster, W.W., and D.F. Hoffmeister. 1954. Life history of the golden mouse, *Peromyscus nuttalli*, in Kentucky. Journal of Mammalogy 35:16–27.

Hall, E.R. 1981. Mammals of North America. 2nd ed., 2 volumes. John Wiley & Sons, New York, New York.

Hall, E.R., and K.R. Kelson, 1959. The mammals of North America, 2 volumes. Ronald Press, New York, New York.

Harlan, R. 1832. Description of a new species of quadruped of the genus *Arvicola*. Monthly American Journal of Geology and Natural Science 1831–1832: 446–447.

Hirth, H.F. 1960. The spermatozoa of some North American bats and rodents. Journal of Morphology 106:77–83.

Hooper, E.T. 1958. The male phallus in mice of the genus *Peromyscus*. Miscellaneous Publication 105, Pages 1–24. Museum of Zoology, University of Michigan, Ann Arbor, Michigan.

Hooper, E.T. 1968. Classification. Pages 27–74 *in* J.A. King, editor. Biology of *Peromyscus* (Rodentia). American Society of Mammalogists Special Publication 2.

Hooper, E.T., and G.G. Musser, 1964. Notes on classification of the rodent genus *Peromyscus*. Occasional Paper 635, Pages 1–13. Museum of Zoology, University of Michigan, Ann Arbor, Michigan.

Howell, A.H. 1921. A biological survey of Alabama. North American Fauna 45:1–88.

Howell, A.H. 1939. Description of a new subspecies of the golden mouse. Journal of Mammalogy 20:498.

Jansa, S., and M. Weksler. 2004. Phylogeny of murid rodents: relationships within and among major lineages as determined by IRBP gene sequences. Molecular Phylogenetics and Evolution 31:256–276.

Komarek, E.V. 1939. A progress report on southeastern mammal studies. Journal of Mammalogy 20:292–299.

Knuth, B.A., and G.W. Barrett. 1984. A comparative study of resource partitioning between *Ochrotomys nuttalli* and *Peromyscus leucopus*. Journal of Mammalogy 65:576–583.

Layne, J.N. 1958. Notes on mammals of southern Illinois. American Midland Naturalist 60:219–254.

Layne, J.N. 1960. The growth and development of young golden mice, *Ochrotomys nuttalli*. Quarterly Journal of the Florida Academy of Science 23:36–58.

Linzey, D.W. 1968. An ecological study of the golden mouse, *Ochrotomys nuttalli*, in the Great Smoky Mountains National Park. American Midland Naturalist 79:320–345.

Linzey, D.W. 1995. Mammals of the Great Smoky Mountains National Park—1995 update. Journal of the Elisha Mitchell Scientific Society 84:384–414.

Linzey, D.W. 1998. The mammals of Virginia. The McDonald & Woodward Publishing Co., Blacksburg, Virginia.

Linzey, D.W., and A.V. Linzey, 1967a. Maturational and seasonal molts in the golden mouse, *Ochrotomys nuttalli*. Journal of Mammalogy 48:236–241.

Linzey, D.W., and A.V. Linzey. 1967b. Growth and development of the golden mouse, *Ochrotomys nuttalli nuttalli*. Journal of Mammalogy 48:445–458.

Linzey, D.W., and R.L. Packard. 1977. *Ochrotomys nuttalli*. Mammalian Species 75:1–6.

Linzey, D.W., M.J. Harvey, E.B. Pivorun, and C.B. Brecht. 2002. Significant new mammal records from the Great Smoky Mountains National Park, Tennessee–North Carolina. Journal of the North Carolina Academy of Science 118:91–96.

Manville, R.H. 1961. The entepicondylar foramen and *Ochrotomys*. Journal of Mammalogy 42:103–104.

Marten Y., G. Gerlach, C. Schlotterer, and A. Meyer. 2000. Molecular phylogeny of European muroid rodents based on complete cytochrome *b* sequences. Molecular Phylogenetics and Evolution 16:37–47.

McCarley, W.H. 1958. Ecology, behavior and population dynamics of *Peromyscus nuttalli* in eastern Texas. Texas Journal of Science 10:147–171.

McCarley, W.H. 1959. The effect of flooding on a marked population of *Peromyscus*. Journal of Mammalogy 40:57–63.

Michaux, J., A. Reyes, and F. Catzeflis. 2001. Evolutionary history of the most spe-ciose mammals: molecular phylogeny of muroid rodents. Molecular Biology and Evolution 18:2017–2031.

Moore, J.C. 1946. Mammals from Welaka, Putnam County, Florida. Journal of Mammalogy 27:49–59.

Morzillo, A.T., G.A. Feldhamer, and M.C. Nicholson. 2003. Home range and nest use of the golden mouse (*Ochrotomys nuttalli*) in southern Illinois. Journal of Mammalogy 84:553–560.

Musser, G.G., and M.D. Carleton. 1993. Family Muridae. Pages 501–755 *in* D.E. Wilson and D.M. Reeder, editors. Mammal species of the world: A taxonomic and geographic reference, 2nd ed. Smithsonian Institution Press, Washington, DC.

Musser, G.G., and M.D. Carleton. 2005. Superfamily Muroidea. Pages 894–1531 *in* D.E. Wilson and D.M. Reeder, editors. Mammal species of the world: A taxonomic and geographic reference, 3rd ed., 2 volumes. Johns Hopkins University Press, Baltimore, Maryland.

Osgood, W.H. 1909. Revision of the mice of the genus *Peromyscus*. North American Fauna 28:1–285.

Packard, R.L. 1969. Taxonomic review of the golden mouse, *Ochrotomys nuttalli*. Miscellaneous Publication 51, Pages 373–406. University of Kansas Museum of Natural History, Lawrence, Kansas.

Patton, J.C., R.J. Baker, and J.C. Avise, 1981. Phenetic and cladistic analyses of biochemical evolution in peromyscine rodents. Pages 288–308 *in* M.H. Smith and J. Joule, editors. Mammalian population genetics. University of Georgia Press, Athens, Georgia.

Patton, J.L., and T.C. Hsu. 1967. Chromosomes of the golden mouse, *Peromyscus* (*Ochrotomys*) *nuttalli* (Harlan). Journal of Mammalogy 48:637–639.

Pearson, P.G. 1953. A field study of *Peromyscus* populations in Gulf Hammock, Florida. Ecology 34:199–207.

Rinker, G.C. 1960. The entepicondylar foramen in *Peromyscus*. Journal of Mammalogy 41:276.

Seagle, S.W. 1985. Competition and coexistence of small mammals in an east Tennessee pine plantation. American Midland Naturalist 114:272–282.

Shadowen, H.E. 1963. A live trap study of small mammals in Louisiana. Journal of Mammalogy 44:103–108.

Strecker, J.K., and W.J. Williams.1929. Mammal notes from Sulphur River, Bowie County, Texas. Journal of Mammalogy 10:259.

True, F.W. 1885. A provisional list of the mammals of North and Central America, and the West Indian Islands. Proceedings of the United States National Museum 8:587–611.

Whitaker, J.O., Jr. 1970. The biological subspecies: An adjunct to the biological species. The Biologist 52:12–15.

Whitaker, J.O. Jr., and W.J. Hamilton, Jr. 1998. Mammals of the Eastern United States. Cornell University Press, Ithaca, New York.

Section 2
Levels of Organization

3
Population Ecology of the Golden Mouse

Robert K. Rose

The golden mouse has low variability in niche configuration, occurs in low abundance even at its optimal site, and is highly susceptible to influence by external or successional habitat alteration. (Dueser and Shugart 1979:115)

An understanding of the population dynamics of a species requires knowledge of the major life-history parameters of a population, including age at maturity, distribution of age classes, and lifetime reproductive contribution of the sexes, sex ratio, length of the breeding season, mean litter size, rates of growth and survival, and life span. Because few long-term studies have been conducted with *Ochrotomys nuttalli* as the focal species of investigation, only fragmentary information is available for many population parameters. As importantly, densities of golden mice often are low, making them difficult to evaluate statistically. Little has been published on age at maturity for golden mice, lifetime reproductive success, or the distribution of age classes in nature. Nevertheless, even early studies provide some useful information focusing on the natural history of this species (e.g., Linzey 1968, McCarley 1958). In this chapter, I summarize studies in which information on one or more parameter(s) is presented, standardize the results as much as possible, and attempt to uncover patterns for populations in one region (e.g., Kentucky, Illinois, Tennessee) to compare with populations from another region (e.g., Florida, Georgia, Texas).

Breeding Season

Reproduction is crucial to sustain any population, and the potential for population increase depends on sex ratio, mating system, age structure, and reproductive behaviors. Additionally, physical factors of the environment must provide resources such as nesting materials, food to support pregnancy and lactation, and a range of favorable temperatures, among others. In north temperate environments, small mammals have a seasonal duration of a few months in which a large proportion of adults is breeding, followed by a nonbreeding season (usually the winter). However, the southeastern

United States, in which the golden mouse is distributed, has locations such as peninsular Florida and the Gulf Coast in which winter weather scarcely exists compared to the northern parts of the distribution in southern Illinois or montane locations in Kentucky, Tennessee, and North Carolina. The golden mouse has adapted to the varied environments in different regions by showing the usual spring–autumn breeding season in most locations. In some years and locations, however, southern populations breed during the winter months and avoid breeding during some summer months (e.g., McCarley 1958, Pearson 1953). Thus, golden mice show the same plasticity in their breeding patterns as is seen with the placement of their nests, usually arboreal but sometimes underground (see Chapter 9 of this volume).

The length of the breeding season and other reproductive details are most accurately determined by necropsying a sample of males and females each month. Such samples reveal the proportions of fertile males and pregnant females, embryo counts (= litter sizes), as well as the body masses and lengths (surrogates for age) of the breeders and nonbreeders. Because such thorough year-long studies have not been published for any population of *O. nuttalli*, other and more fragmentary information is used, such as litters born in traps or captivity, backdating to determine the time of birth of half-grown young in nest boxes or traps, embryos counted during necropsy, or the changing reproductive indices of animals captured and released over the course of a year.

In the Smoky Mountains of Tennessee at elevations of 1837–2722 ft (560–830 m), the breeding season extends from mid-March to early October (Linzey 1968). Peaks of breeding are in late spring and early autumn, as determined by the appearance of juveniles in the trappable population, the presence of embryos and sperm in necropsied adults, and from litters born in captivity. A breeding season of similar length is reported for northern Kentucky, based on the presence of preg nant females (Goodpaster and Hoffmeister 1954).

By contrast, the breeding season often differs in southern locations. For example, Ivey (1949), in eastern Florida, found a female with suckling young and four embryos on 3 November, a female with young about 1.5 months old on 21 December, and a female with young about 1 week old on 21 December. This information indicates breeding during October to December, and perhaps longer in the winter.

Pearson (1953) found peak density in the January–May period in central Florida, suggestive that a population increase via reproduction was occurring during early winter. Also in central Florida, Layne (1960) recorded one litter born in January and four litters born in July, pregnant and lactating females in July, September, and November, and a female with newborn young in a nest on 2 March. Thus, Layne suggested an 8- or 9-month breeding season. In a population study conducted in eastern Texas, McCarley (1958) reported that in January and February, most mature mice were in breeding condition, and winter breeding is also supported by the appearance of juveniles and by the attainment of the highest population densities during the winter months. McCarley (1958) stated that unlike northern populations, southern populations of golden mice breed during the winter. These studies of southern populations of golden mice support his contention.

Litter Size

Litter size is approximately equivalent to embryo counts, assuming that all embryos survive until parturition. Both kinds of information are useful in determining when breeding starts and ends, and potential population increase. Using all previous information on litter size for golden mice in Florida, Layne (1960) calculated a mean litter size of 2.7 with a modal value of 2, not different from what Goodpaster and Hoffmeister (1954) and McCarley (1958) reported for Kentucky and Texas populations, respectively. In an intensive evaluation of litter size for this species, Linzey (1968) reported a mean litter size of 2.65 for 85 litters from Tennessee, with mean spring litters smaller (2.4) than those born in September to November (3.1). Blus (1966a) summarized published information on embryo counts and litter sizes and concluded that mean litter size (3.11 ± 0.10) in northern populations (Illinois, Kentucky, North Carolina, and Tennessee) was significantly greater ($p < 0.001$) than that (2.47 ± 0.11) of southern populations (Georgia, Florida, and Texas). Thus, golden mice seemingly follow the pattern of many other species of small mammals with regard to litter size; northern populations have larger mean litter sizes than southern populations (Lord 1960).

Growth and Development

Like many species of small mammals, golden mice have a brief gestation period, 25–26 days (Linzey and Linzey 1967a), followed by a period of nursing and development in the nest. The male plays no observed role in the rearing of young. Newborn golden mice are naked, blind, and helpless, but their rate of growth and development is rapid, so that independence (weaning) is achieved by 3 weeks of age and sexual maturity follows within a few weeks. Spring-born animals breed during the year of their birth, but autumn-born ones might not, especially in more northern populations. The dynamics of this progression to entry into the breeding population remain to be explored for southern populations, which breed in winter. Newborns weigh, 2.4–2.6 g, but their rate of growth is 12.6 percent per day for the first week (Layne 1960), by which time they weigh about 5 g (Figure 3.1). Growth rate slows thereafter, but at 14 days the young weigh 7.5 g and at 3 weeks they weigh 10.5 g, approximately half of the adult weight (Linzey and Linzey 1967a).

Physical development occurs in a predictable manner. By day 2, hairs on the back are darkening and extending to the tail, and by day 3, the sutures of the skull have closed (Layne 1960). By day 4, young are able to right themselves easily and are becoming more coordinated in their movements. By day 6, littermates were considered "agile" (Wallace 1969), and their lower incisors erupt (Linzey and Linzey 1967a). Upper incisors erupt a day later on average, and soon the adult proportion of the lower incisor, being twice the length of the upper, is achieved. Eyes open between days 11 and 14 (mean of 12.7 days [Linzey and

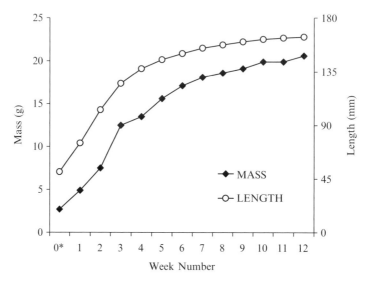

FIGURE 3.1. Growth trajectories of body mass and total length for golden mice for the first 12 weeks of life, based on the laboratory study of Linzey and Linzey (1967a). The asterisk (*) represents the measurements at birth.

Linzey 1967a]), after which the young move about freely (deRageot 1964). The juvenile pelage is fully developed by day 14 (Layne 1960). By day 15, squeaking gives way to chattering, nursing is infrequent, and adult eye shape is attained (Wallace 1969). Some young are weaned at day 17 or 18, most by day 21, and all by day 24 (Linzey and Linzey 1967a, Wallace 1969). The postjuvenile molt (leading to adult pelage) starts at week 4 or 5 and usually is completed in 10–14 days (Layne 1960). At this time, youngsters weigh about 19 g, or nearly adult size (Linzey and Linzey 1967a). Young females become fertile and typically enter the breeding population earlier than males.

Spring-born litters gained weight faster than autumn-born litters, and although their linear body measurements were reversed, these differences were slight (Linzey and Linzey 1967a). Layne (1960) concluded that developing young golden mice are superior in many ways to those of syntopic *Peromyscus* species; they are relatively larger at birth, develop more rapidly, and acquire adaptive behaviors sooner.

Because females suspend molting during pregnancy, an indication of seasonal breeding sometimes can be gleaned from detailed molting information. For example, Linzey and Linzey (1967b) collected 10 golden mice between 12 and 17 December in the Smoky Mountains of Tennessee, of which 6 were molting and 4 already were in winter pelage, indicating that breeding had just ended. By contrast, all wild golden mice observed between 26 March and 1 April were still in winter pelage but had molted by 15 June; this observation indicates that litters likely were born into this population later in April.

Density

Population density—the number of individuals per unit area—depends on the rate at which animals enter the population through recruitment by both reproduction and immigration and the rate at which animals disappear, via either emigration or death. Increases in density usually coincide with the entry of young into the population, but little is known about the role of immigrants in contributing to density or of the site fidelity or dispersal of young golden mice. Even less is known about gross mortality (i.e., losses via death or emigration). Density is also a measure of the health and vigor of a population; high densities suggest successful and probably persisting populations as well as quality habitat, whereas low densities more likely indicate newly established or nearly extirpated populations and perhaps marginal habitat. The presence of potentially competing arboreal species, usually in the genus *Peromyscus,* also often is considered when evaluating density of golden mice. Because golden mice often have patchy distributions, the high variability in estimates of density among geographic populations (Table 3.1) is due in part to whether investigators establish their study grids in optimal habitat for golden mice, or at random within a large homogeneous study area, or the season or duration of the study.

The best estimates of population density are obtained when measured grids are used to study a population using capture–mark–release (CMR) methods over a period of months and years. Few such studies have been published for golden mice. More often, the studies using CMR methods last one summer or parts of two seasons. In addition, instantaneous estimates of population density can be determined when intensive grid trapping, whether with live traps or snap traps, is concentrated within such a brief period as to reduce the confounding effects of birth and death. In such instances, I have used the inclusive boundary strip method (Stickel 1954) to determine the effective area of trapping. This method adds a perimeter of half the trap interval to the enclosed area of the grid. Thus, if the grid is 10×10 with 10-m intervals, I have added a 5-m boundary strip to the perimeter in determining the effective area of trapping, which, in this example, is 1.0 ha.

Most studies of golden mice have been conducted in deciduous floodplain or mesic upland forests (e.g., Linzey 1968, McCarley 1958, Schmid-Holmes and Drickamer 2001), but sometimes populations have been studied in pine forests or even pine plantations (Mengak and Guynn 2003, Perry and Thill 2005). Whatever the forest type, dense thickets with tangles of vines seem to be required, as noted throughout this volume. The golden mouse is arboreal but spends a variable (and largely unknown) percentage of time on the ground. One or more other arboreal small mammal(s), usually in the genus *Peromyscus*, is syntopic (live in the same forest community as the golden mouse); all are primarily nocturnal.

Linzey (1968) conducted one of the longest field studies of golden mice using CMR methods. He trapped from June 1964 to August 1966 in the Great Smoky Mountains (Tennessee) National Park and reported a maximum density of 1.5 residents/ha in September 1964 and a low density of 0.1 residents/ha in

TABLE 3.1. Population densities of published studies of *O. nuttalli*.

Number/ha	Location	Habitat	Reference
1.5 (during 1964)	Tennessee	Early successional forest with vines	Linzey 1968
0.4 (during 1965)	"	"	"
4.7 (first winter)	Eastern Texas	Pine/oak forests	McCarley 1958
6.2 (second winter)	"	"	"
1.1 (preburn)	Northern Louisiana	Pine forests (Loblolly-shortleaf)	Shadowen 1963
0.4 (postburn)	"	"	"
0.3	Central Florida	Evergreen mesic forest	Pearson 1953
2.0–74.1	Southern Illinois	Floodplain forest	Blus 1966b
6.5	Central Florida	Pine/oak/palmetto	Frank and Layne 1992
<3	Eastern Tennessee	Mixed deciduous/coniferous	Dueser and Shugart 1979
12 (summer)	Eastern Tennessee	Cedar, pine, oak with vines	Kitchings and Levy 1981
15 (autumn)	Eastern Tennessee	"	"
17.2 (summer)	Eastern Tennessee	Cedar glade forest: red cedar/shortleaf pine/ sweet gum	Seagle 1985a
0 (summer)	Eastern Tennessee	Deciduous forest: chestnut oak/white oak/beech	"
7.8 (with *P. leucopus*)	Eastern Tennessee	Pine plantation	Seagle 1985b
18.8 (without *P. leucopus*)	"	"	"
29–63	Georgia	Deciduous riparian forest	Christopher and Barrett 2006
0.3	Southern Illinois	Pine plantations	Feldhamer and Maycroft 1992
23.9	Central Florida	Shrub habitat; sand pine, closed canopy	Packer and Layne 1991
2.5	Central Florida	Sandhill habitat; slash pine, open canopy	"
1.6	South Carolina	Mesic hardwood forest	Faust et al. 1971
13.4 (summer)	South Carolina	Mesic hardwood forest	Smith et al. 1971
3.9 (6 summers)	South Carolina	Hardwood forest	Smith et al. 1974
<2	Eastern Virginia	Hardwood forest	Rose and Walke 1988

Note: Estimates vary widely in part because of differences in methods.

September 1965, a year of reproductive failure (Table 3.1). McCarley (1958), who used a combination of live traps and nest boxes in a 29-month study of golden mouse populations of two large floodplain grids and one large upland (pine/oak) grid in eastern Texas, recorded the highest densities at the end of winter, (4.7 residents/ha in the winter of 1955–1956 and, 6.2 residents/ha in 1956–1957) on one plot, with lesser densities on the other two plots. The period of peak density was January to May, coinciding with the appearance of young during the winter breeding season.

Shadowen (1963) live-trapped *O. nuttalli* and the cotton mouse (*P. gossypinus*) for 28 months on two large grids in loblolly (*Pinus taeda*) and shortleaf (*Pinus echinata*) pine forests in northern Louisiana. The highest densities were 1.1 residents/ha for golden mice and 0.5 residents/ha for cotton mice. Part way through the study, one of the areas was burned and the responses of the species were compared. After the burn, the population of golden mice declined by 65 percent, whereas that of the cotton mouse increased by 155 percent. This change in proportions, reflecting the loss of habitat structure after the fire, was highly significant ($\chi^2 = 149.14$, 3 *df*, $p < 0.001$).

Pearson (1953), using CMR methods in a 1-year study on a 3.64-ha site dominated by evergreen mesic forest in central Florida, caught 18 golden mice and 89 cotton mice (1:5 ratio). The highest density (0.3 residents/ha) of golden mice was seen in May at the end of the breeding season. By contrast, the cotton mice had a peak density (1.4 residents/ha) in October. He suggested that golden mice are reluctant to enter traps, unlike most species of *Peromyscus*. However, once individual golden mice were trapped, they entered traps (four recaptures/animal) as readily as cotton mice (three recaptures/animal).

In another year-long live trapping study, this one on five plots in southern Illinois, Blus (1966b) reported that most densities were less than 1 resident/ha (Table 3.1). However, on one grid in January, he caught 30 animals per acre or 74.1 residents per hectare (Table 3.1). Frank and Layne (1992) used June–September and January–March live trapping on a 2.8-ha grid to determine similar densities (6.5 residents/ha) in summer and winter populations of golden mice in slash pine (*P. elliottii*)/turkey oak (*Quercus laevis*)/saw palmetto (*Serenoa repens*) habitat in central Florida. *O. nuttalli* outnumbered *P. gossypinus* 17:6 in summer, but in winter live trappingthe approximate 3:1 ratios were reversed, with cotton mice outnumbering golden mice 51:17.

Kitchings and Levy (1981), who used two 10-day periods of grid trapping in summer and a longer period in autumn in a forest at the Oak Ridge National Environmental Research Park (ORNERP) in eastern Tennessee, also found that seasonal densities of golden mice were more constant than those of the syntopic species of *Peromyscus*. Densities of *Ochrotomys* were 12 residents/ha in summer and 15 residents/ha in autumn, whereas those of *P. leucopus* were 15.6 residents/ha in summer and 37.5 residents/ha in autumn. These last two studies perhaps indicate a greater degree of intrinsic population regulation for golden mice than for either *Peromyscus* species.

Dueser and Shugart (1979:115), who measured habitat variables to examine niche pattern in a mixed deciduous/coniferous forest at ORNERP, noted that *Ochrotomys* occupies a position apart from other small mammals, "has low variability in niche configuration, occurs in low abundance even at its optimal site, and is highly susceptible to influence by external or successional habitat alteration." Thus, even in prime habitat, densities often were low. They caught less than 3 residents/ha in trapping conducted during 4 summer months.

In a live trapping study conducted at ORNERP during the summer, Seagle (1985a) recorded a density of 17.2 golden mice/ha (Table 3.1) and 25.0 white-footed mice (*P. leucopus*)/ha on a grid in a cedar glade. However, in deciduous forest, he captured no golden mice and 35.9 *P. leucopus*/ha.

The preceding summer, Seagle (1985b) had studied syntopic *O. nuttalli* and *P. leucopus* in a loblolly pine plantation at ORNERP. Using a pair of 1.6-ac (0.64-ha) plots and Sherman live traps, he marked animals for 6 weeks, then removed the *P. leucopus* from one plot, testing for density compensation by *Ochrotomys*. Densities of golden mice did increase in the absence of *P. leucopus*, from 7.8 residents/ha to 18.8 residents/ha, primarily by immigration of animals during the last 7 weeks of trapping. During this interval, densities of golden mice on the control grid decreased from 15.6 to 10.9 resients/ha.

In another study in which the density response by golden mice to the removal of a potentially competing *Peromyscus* species was examined, Christopher and Barrett (2006), after making density determinations, removed *P. leucopus* but not other species from 0.21-ha experimental plots in Georgia floodplain and upland deciduous forests. Before removal, the mean abundance of *P. leucopus* was 25 individuals per grid and the mean maximum abundance of *O. nuttalli* was 15 per grid. Although densities of *Ochrotomys* did not increase significantly in the absence of *P. leucopus*, they were exceedingly high, from 63 to 92 residents/ha on the multiple treatment and control plots. The authors set traps at ground level and also at 1.5 and 4.5 m above ground. Golden mice used all elevations equally in the presence of *P. leucopus*, but after the white-footed mice had been removed, golden mice used the 4.5-m traps less and the other two elevations equally. Thus, despite no density compensatory response, golden mice altered activity somewhat, becoming less arboreal in the absence of *P. leucopus*.

While livetrapping during the summer on 1.6-ac (0.64-ha) grids in 21 pine plantations in southern Illinois, Feldhamer and Maycroft (1992) caught 45 *O. nuttalli* on 13 sites; white-footed mice were always present ($n = 96$) on the same sites. The density of golden mice was 0.3 residents/ha and that of white-footed mice was slightly more than twice that value. Five other sites had only *P. leucopus* and three sites had neither species. In a study in which golden mice were sought in 18 forested sites in 15 counties in southern Illinois, Feldhamer and Paine (1987) captured 38 *O. nuttalli* and 370 *P. leucopus*. The latter species was 10 times more abundant on these sites.

Packer and Layne (1991), using live traps during January–April, reported densities of 17.6 golden mice/ha in dense scrub habitat but less than 2.0 golden mice/ha in the more open sandhill habitat in central Florida. In scrub habitat, the 67 golden mice were more than double all other species, including cotton mice (25), old-field mice (*P. polionotus*) (0), Florida mice (*Podomys* [formerly *Peromyscus*] *floridanus*) (2), and hispid cotton rats (*Sigmodon hispidus*) (1). By contrast, on the sandhill habitat, the 7 golden mice comprised 5 percent of total individuals, which included 30 *P. gossypinus*, 24 *P. polionotus*, 61 *Podomys floridanus*, and 13 *S. hispidus*.

During 63 consecutive days of mid-summer live trapping on a 16.4-ha grid in lowland mesic hardwood forest at the Savanna River Ecology Laboratory (SREL) near Aiken, South Carolina, Faust et al. (1971) recorded densities of acre (1.6/ha) for golden mice and acre (2.7/ha) for cotton mice. In another study conducted at SREL, Smith et al. (1971) used snap traps in a lowland mesic hardwood forest to determine population densities of small mammals. Initially during 18 consecutive days of trapping on the grid, and later by trapping on assessment lines radiating from the grid, they captured 87 golden mice and 56 cotton mice (and other small mammal species), for density estimates of 13.4/ha for golden mice and 8.4/ha for cotton mice. This is another example in which golden mice were numerically dominant to another arboreal rodent of similar body mass and life history.

In yet another study at SREL, this one conducted on a large grid in mature cove hardwood forest (Smith et al. 1974), the average number of golden mice (24.7 ± 5.3) exceeded that of cotton mice (21.2 ± 4.4). Converted to density, the mean number of golden mice was 3.9/ha for six consecutive summers of study. Thus, the four studies in which density of golden mice was greater than that of *P. gossypinus* were Smith et al. (1971) and Smith et al. (1974), both conducted in South Carolina, Shadowen (1963) in Louisiana, and Packer and Layne (1991) in Florida.

In another study in which golden mice were the numerically dominant small mammal (but for which no density estimates could be made), Miller et al. (2004) used transects of snap traps to evaluate the small mammal community in stream-side management zones (SMZs) in intensively managed loblolly pine plantations located in Arkansas. *O. nuttalli* comprised 36.9 percent of the 1701 total captures, *Blarina carolinensis* (southern short-tailed shrew) comprised 28.9 percent, and 4 *Peromyscus* species combined comprised 21.9 percent. SMZs were dominated by hardwoods; for those (> 60 m) wide, > 70 percent of golden mice were captured in traps at the boundary between pine plantation and SMZ.

In the only example in which golden mice were more numerous than white-footed mice, Dolan and Rose (2007), using live traps followed by pitfall traps in seasonal trapping in different-aged loblolly pine plantations in eastern Virginia, caught highly varying proportions of both species. On three 1-year-old stands, *O. nuttalli* was absent, but 72 ($\bar{X}$ = 24) *P. leucopus* were caught. Means for the other replicated stands (*O.nuttalli*:*P. leucopus*) were as follows: 8-year-old stands = 6:0, 18-year-old stands = 4.5:0.5, and 24-year-old stands = 2.7:4.6. Commercial thinning in two of the three oldest pine stands likely added the shrubby structure to enable golden mice to persist; unthinned stands of this age have little undergrowth below the canopy. Loeb (1999) also noted that *O. nuttalli* was captured only in salvaged (regenerating) plots following tornado destruction of longleaf pine (*P. palustris*) stands in the Upper Coastal Plain of South Carolina.

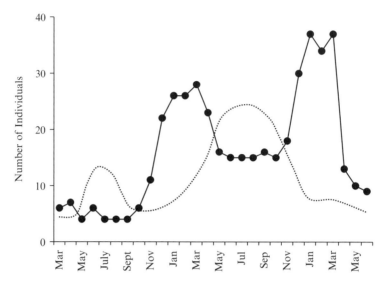

FIGURE 3.2. Patterns of abundance for populations of golden mice from the South (black dots = eastern Texas; based on McCarley 1958) and a hypothetical population from the North (dashed line). Populations from the South attain highest densities in winter/spring, whereas those from the North in summer/autumn.

Finally, information is available from two other studies conducted in Virginia, near the northern limit of distribution. Using tree-mounted nest boxes on large grids in the seasonally flooded hardwood forest in the Dismal Swamp of Virginia, Rose and Walke (1988) recorded golden mouse density of less than 2/ha. *Ochrotomys* was absent on 2 of the 4 grids, and, overall, *P. leucopus* outnumbered golden mice by 36:9. In a study using live traps distributed in quality habitat at 3 sites in southwestern Virginia, Wilder and Fisher (1972) caught 30 golden mice and 211 white-footed mice. No density estimate is possible, but in this study *P. leucopus* outnumbered *O. nuttalli* 7:1.

Figure 3.2 shows the hypothetical differences between the peak densities of northern populations, with peak densities in late autumn, compared to those of southern populations, in which peak densities sometimes are achieved in late winter or early spring.

Sex Ratio

When the sex ratio of golden mice is examined, parity is the usual result. For example, Goodpaster and Hoffmeister (1954) found 22 males and 24 females in nests during the winter. Likewise, Linzey and Linzey (1967a) recorded 105 males and 97 females in their studies of litters. Significant deviations from parity, however, include 20 males and 6 females ($\chi^2 = 7.54$, 1 df, $p < 0.01$) taken with snap traps in South Carolina (Faust et al. 1971), and 94 males and 58 females ($\chi^2 = 8.53$, 1 df, $p < 0.001$) reported by Linzey (1968) in his field study in the Great Smoky Mountains. The ≥ 50 percent proportion of adult males every month (Linzey 1968) is suggestive of differential mortality.

Age Distribution

Frank and Layne (1992), using CMR methods in Florida, reported that only adult golden mice were present in summer, but the winter population had slightly more subadults than adults, results that support winter breeding. McCarley (1958), who presented proportions of adults:immatures for both sexes for each month of his study, reported high proportions of adults, often 100 percent, during the May–October period. The percentages of immatures were highly variable during November–April. These proportions also support winter breeding in this species. Linzey (1968) found the highest proportions of immatures in June, July, and December and mostly adults in other months.

Patterns of Dispersion

Dispersion refers to the pattern in which organisms are distributed in space, whether uniformly (as rows of trees planted in a pine plantation or pieces positioned on a chess board), randomly (no pattern), or clumped (aggregated). Due in part to their habitat requirements, golden mice usually are found in clusters, whether in dense viny thickets in forests or along forest edges. Many investigators stated that when one arboreal nest is found, others likely will be found nearby. These clusters of nests often are separated by large areas of seemingly similar habitat in which no arboreal nests or golden mice can be found (e.g., Ivey 1949, Pearson 1953). Sometimes investigators measured habitat complexity, but usually the assessments were subjective.

For example, McCarley (1958) categorized the condition of underbrush in a pine-oak upland forest in eastern Texas as dense, intermediate, or sparse. Traps and nest boxes at locations with dense vegetation had significantly ($\chi^2 = 26.78$, 2 df, $p < 0.001$) more captures than in less vegetated trapping locations. He concluded that dispersion in *O. nuttallii* was related to, and perhaps regulated by, amounts of brushy and dense vegetation. McCarley (1958), who frequently observed more than one golden mouse in a nest box, found that ground and tree nest boxes were used equally ($\chi^2 = 0.96$, 1 df, $p > 0.50$). Ivey (1949:160), who found greater than 1000 nests from 1939 to 1942 in eastern Florida, reported that "golden mice appear to live in rather loosely knit communities consisting of 3-4 occupied nests. When one nest is discovered, there are likely to be others, while in large tracts of similar country it is impossible to find a single nest." This suggests a nonrandom distribution of groups of (perhaps related) golden mice. Nests some-times had more than one inhabitant, including pairs, but males were never found in nests with young nor were young of two ages found together.

Pearson (1953:206), in central Florida, reported that except for one male and one female, all other resident golden mice "were clustered in a restricted area of about one acre (0.4 ha) of dense shrubby habitat on the mesic ridge." This is another example of the nonrandom dispersion of golden mice in a diverse landscape.

At ORNERP near Knoxville, golden mice showed a strong affinity for an open area at the margins of a swamp in the summer but used an area dominated by cedars and avoided the swamp in autumn. This pattern of movement suggests that areas of occupied habitat might change with changing availability of resources (Kitchings and Levy 1981). Thus, patterns of dispersion might be less static from season to season than previously believed.

In addition to residents (some caught up to 40 times), Linzey (1968) reported the presence of a small number ("3-4 each month") of transients—animals trapped once or twice. This suggests highly sedentary as well as clumped groups of residents.

The high proportion of overlapping home ranges of both sexes suggests a gre-garious nature for the 19 radiocollared golden mice monitored by Morzillo et al. (2003) in southern Illinois. The home ranges of 9 of 19 mice overlapped almost completely, and nest sharing occurred among all 5 females in this group. However, no nest sharing was observed among the 4 males nor among any hetero-sexual groups.

Blus (1966b) found 87 hollow globular nests, mostly unoccupied, located an average of 13.1 m apart. However, some nests were isolated from others by dis-tances of > 200 m, suggesting a random distribution within a cluster of nests. Distances between two nests occupied on the same day were 42.6 m and 54.9 m.

Dunaway (1955), by following paint-marked individuals during late autumn in Tennessee, found three females often living in the same nest. In yet another exam-ple of a group of golden mice living in close proximity, Eads and Brown (1953) collected five adults and six juveniles at the base of a rotten stump in a bottom-land of deciduous forest dominated by sweet gum (*Liquidambar styraciflua*), bay (*Gordonia lasianthus*), oak (*Quercus* spp.), and holly (*Ilex opaca*), with dense undergrowth of cane (*Arundinaria gigantea*), greenbrier (*Smilax* spp.), and black-berry (*Rubus* spp.) located near Tuscaloosa, Alabama. The juveniles must have come from two or more litters, because litters of more than four are unknown in *O. nuttalli*. This finding again indicates clumped distribution. In yet another example of a group of golden mice living in close proximity, Goodpaster and Hoffmeister (1954) found multiple males and females in the same nests in Kentucky, mostly in the winter. The largest number was three of each sex from a nest on 24 November 1951.

Finally, in another study in southern Illinois, Andrews (1963) found eight golden mice clustered in "close proximity" in one drainageway of an eroded upland old field. Although absent in that drainageway, white-footed mice occu-pied all other drainageways.

Home Range

Home range, as defined by Burt (1943:351), is "that area traversed by the individual animal in its normal activities of food gathering, mating, and caring for young." Home range size, often assessed by determining the enclosed area using captures at marginal locations, is more accurately estimated by using locations determined by radiotelemetry. Sometimes home range is estimated by a linear measurement, such as longest linear distance moved or the longest distance moved from a nest or trap. The term "territory," an area defended, has not been used by any investigator studying golden mice. I used the inclusive boundary strip method of Stickel (1954) to recalculate home ranges from published studies (Table 3.2).

For 39 males captured 3-40 times, the home range increased with the number of captures (Linzey 1968). For males with ≥ 9 captures, the average home range size was 0.26 ha; females averaged 0.24 ha (Table 3.2). The largest home ranges were 0.63 ha for a male and 0.39 ha for a female. Using the average distance moved between successive captures as another index of home range, Linzey (1968) reported values of 31.4 m in 1964 and 59.3 m in 1965, when densities were much lower. Values for females were 31.7 and 27.1 m, respectively.

Home ranges of males (0.60 ha) also were similar to those of females (0.54 ha) for animals caught ≥ 10 times in eastern Texas (McCarley 1958). McCarley (1958) also found that the area of home ranges was related to the number of captures; the largest home range for a female captured 29 times was 1.26 ha. Shadowen (1963) calculated the home range size for *O. nuttalli* to be 0.53 ha on a control plot (loblolly/shortleaf pines), and 0.43 ha on a burned plot in northern Louisiana. Pruett et al. (2002), using radiotelemetry, reported a mean home range size for male *O. nuttalli* as 0.90 ha and 0.50 ha for females. The home ranges of males and females did not differ significantly in any of these three studies.

Dunaway (1955) reported mean home range size of 0.11 ha for three female golden mice captured from late October to early December in Tennessee. Using a 16.4-ha grid of live traps, Faust et al. (1971) calculated home range sizes of males (0.93 ha) and females (0.38 ha) in South Carolina, using the inclusive boundary strip method. They also determined that the average distances between successive captures were 77.6 m and 37.6 m, respectively.

Some investigators believed that golden mice have small home ranges based on knowledge of their environment. For example, Goodpaster and Hoffmeister (1954:20) stated that "although we have no figures, it is firmly believed that the home range is small, and that nightly forays are made short distances from the nest. It is necessary to set traps close to occupied nests to take specimens." deRageot (1964) agreed, saying that the home range size is small because animals are trapped close to their nest sites.

Morzillo et al. (2003), using radiotelemetry to find arboreal and ground nests, determined the home ranges of golden mice in southern Illinois. Whether using minimum convex polygon (MCP) method (0.53 ha for males, 0.37 ha for females)

TABLE 3.2 Estimates of home ranges (HR) in hectares from published studies of *Ochrotomys*.

HR (male/female)	Criteria	Method	Estimator	Location	Reference
0.26/0.24 in 1964	9 Captures	Live traps	Exclusive b. strip	Tennessee	Linzey 1968
−/0.24 in 1965	9 Captures	Live traps	Exclusive b. strip	Tennessee	Linzey 1968
0.60/0.54	≥10 Captures	Live traps	Inclusive b. strip	Texas	McCarley 1958
−/0.11	15–30 Captures	Live traps	Minimum area	Tennessee	Dunaway 1955
0.93/0.38	≥4 Captures	Trapping	Inclusive b. strip	S. Carolina	Faust et al. 1971
0.53 (control grid)	≥3 Captures	Live traps	Widest capture points	Louisiana	Shadowen 1963
0.43 (burned grid)	≥3 Captures	Live traps	Widest capture points	Louisiana	Shadowen 1963
0.53/0.37	20–45 Locations	Radiotelemetry	MCP	Illinois	Morzillo et al. 2003
1.34/1.11	20–45 Locations	Radiotelemetry	Kernel	Illinois	Morzillo et al. 2003
0.53/0.61	5–9/10 Captures	Trapping	Inclusive b. strip	Illinois	Blus 1966b
0.13 (both sexes)	4–5 Captures	Live traps	Inclusive b. strip	Arkansas	Redman and Sealander 1958
0.20/0.15	≥5 Captures	Live traps	MCP	Georgia	Pruett et al. 2002
0.90/0.50	5–15 Locations	Radiotelemetry	MCP	Georgia	Pruett et al. 2002

Note: Estimates vary widely in part because of differences in methods.

or average kernel estimate (1.34 ha and 1.11 ha, respectively), home range sizes between sexes did not differ. Nor were there differences seasonally—when the trees had leaves or not. Further, 18 of 19 radiocollared animals had overlapping home ranges, 9 of which overlapped with only 1 other golden mouse. One male–female pair shared a nest every day during observation, but another never shared nests. Among pairs, the percentage of overlap differed significantly for the MCP method but not for kernel estimates. Home range overlaps for female–female pairs were 38 percent (MCP) and 46 percent (kernel), 32–35 percent for male–female pairs, and 21 percent (MCP) and 34 percent (kernel) for male–male pairs. The percentages of home range overlap were not different between seasons. Thus, the results of several studies support the assertion of similar or equal home range areas for males and females.

Blus (1966b), using the greatest linear distance traveled, found that males, caught 4–10 times, traveled from 42.4 to 147.2 m during a period of 5 months. The longest linear distance by a female, caught 15 times in March, was 60.4 m. As with home range calculations, more captures usually translates to greater distances moved. Pearson (1953) did not calculate home ranges for golden mice in central Florida, but he did determine the distances moved from release to recapture points. Golden mice moved less widely than syntopic cotton mice, and 62 percent of recaptures of *O. nuttalli* were at distances < 38 m, perhaps indicating a sedentary nature. Whether using area or linear measurements, the home range of a golden mouse is, as presently understood, approximately equal to the area of two-thirds of a football field.

As noted by Meserve (1977) and Christopher and Barrett (2006), investigators estimating home range in arboreal small mammals should consider not just area (length × width dimensions) but also elevation (height). The realm of golden mice and other arboreal small mammals really is three dimensional.

Longevity

Most species of small mammal have short life spans, measured in weeks or months rather than years. For populations in temperate (northern) locations, autumn-born young usually live longer than spring-born young, but the pattern of longevity is less clear for southern populations of golden mice that breed in the winter. Survival rates of small mammals usually are higher in winter than in other seasons, probably because breeding is suspended. In natural populations, only a tiny proportion (1–3 percent) of small mammals, including golden mice, live long enough to see another season of their birth. It would be interesting to learn whether winter-born cohorts of golden mice in southern populations have different life spans than autumn-born cohorts in northern populations.

During his 28-month live trapping study in the Great Smoky Mountains, Linzey (1968) recaptured many golden mice multiple times. Nineteen mice lived ≥ 5 months, 10 mice lived ≥ 8 months, and 2 mice lived for one full year.

McCarley (1958) stated (but presented no data in support) that adult male golden mice had slightly longer life spans than females in eastern Texas. On one plot, 38 adults had mean minimum life spans of 6.8 months and 39 immatures averaged 6.0 months; on another plot, the values were 6.5 months and 3.1 months, respectively. McCarley (1958) estimated that 15 percent of animals lived 6 months or more, which seems to be a fairly high percentage for a small mammal. The longest field life he observed for a golden mouse was 19 months.

Pearson (1953), working in a mesic evergreen forest of central Florida, found that 6 of 10 resident golden mice lived longer than 3 months after reaching adulthood, and one lived nearly 2.5 years. In loblolly/shortleaf pine forests of northern Louisiana, Shadowen (1963) reported that one golden mouse lived nearly 10 months and six others lived 6 months after their initial captures.

Population Genetics

Little is known regarding the population genetics or metapopulation dynamics of golden mice. However, basic information is known about their chromosomes, including that their karyotypes show no variation in gross morphology of the chromosomes (Engstrom and Bickham 1982, Patton and Hsu 1967). *O. nuttalli* has a diploid number of 52 chromosomes, including 3 pairs of subtelocentrics, 8 pairs of metacentrics, and 14 pairs of acrocentrics of various sizes. The Y-chromosome is a small acrocentric, indistinguishable from the smallest acrocentric autosome. The X-chromosome is slightly smaller than the largest subtelocentric autosome, and its short arm is slightly longer than that of the largest subtelocentric autosome. All members of the genus *Peromyscus* have 48 chromosomes (Patton and Hsu 1967). Engstrom and Bickham (1982), using G- and C-banding techniques, determined that chromosomes 1 and 13 of *Ochrotomys*, which do not seem to be shared with peromyscines, appear to be homologous to chromosomes 1 and 10 of *Sigmodon hispidus*, proposed as their closest relative.

A phylogeographic study of the golden mouse using molecular methods is needed, one that should include its former congeners, *P. gossypinus, P. poliono-tus, P. leucopus*, and *S. hispidus*. Additional future opportunities for research on the golden mouse are noted by Barrett and Feldhamer (Chapter 12 of this volume).

Conclusions

Because *O. nuttalli* often is studied as a secondary rather than a focal species, only a few studies have evaluated its major life-history parameters. Nevertheless, the available information summarized in this chapter provides some tantalizing possibilities and opportunities for further study. For example, some southern populations of golden mice have deviated from the usual spring–autumn breeding season by sometimes breeding primarily during the cooler months and attaining highest densities in late winter or early spring. The frequency and adaptive value of this strategy remain to be determined, and several important research questions (genetic, ecological, physiological, and behavioral) relate to these southern populations. Compared to its close relatives in the genus *Peromyscus*, golden mice have small litter sizes, indicative of relatively longer lives than its former congeners if Lack's (1966) hypothesis that rates of natality have evolved to compensate for rates of mortality is correct. An indication of relatively long life spans also supports this contention of higher rates of survival than for other similar arboreal small mammals living in the same habitat. That golden mice often live in clusters of nests separated from other groups by unoccupied suitable habitat suggests a high degree of relatedness among individuals. Home ranges often are small and are similar between the sexes, perhaps indicating high levels of tolerance for neighbors, which might be relatives. Genetic studies would answer many of these questions.

One of the challenges when studying golden mice is that their specialized habitats of dense understory thickets of tangled vines in a forested matrix often disappear as secondary succession proceeds. Thus, the brushy habitat provided by 8–10-year-old pine trees disappears quickly as the trees grow taller, thereby thinning the volume below the canopy. The same is true, probably at a slower rate, for deciduous forests. Consequently, what is good golden mouse habitat this year might not support a population in 3 or 5 years. Old fields giving way to secondary succession by the Eastern red cedar (*Juniperus virginiana*) might provide the greatest promise as suitable habitat for sustaining populations of golden mice for long-term investigations. Because they exhibit less self-thinning than other conifers, red cedars, as they grow, continue to provide the three-dimensional habitat seemingly required by golden mice. Hopefully, the National Science Foundation will recognize this opportunity by funding a collaborative LTREB grant. In all, much remains to be learned about the population biology of the golden mouse.

Literature Cited

Andrews, G.D. 1963. The golden mouse in southern Illinois. Natural History Miscellanea 179:1–3.

Blus, L.J. 1966a. Relationship between litter size and latitude in the golden mouse. Journal of Mammalogy 47:546–547.

Blus, L.J. 1966b. Some aspects of golden mouse ecology in southern Illinois. Transactions of the Illinois Academy of Science 59:334–341.

Burt, W.H. 1943. Territory and home range concepts as applied to mammals. Journal of Mammalogy 24:346–352.

Christopher, C.C., and G.W. Barrett. 2006. Coexistence of white-footed mice (*Peromyscus leucopus*) and golden mice (*Ochrotomys nuttalli*) in a southeastern forest. Journal of Mammalogy 8:102–107.

deRageot, R.H. 1964. The golden mouse. Virginia Wildlife 25:10–11.

Dolan, J.D., and R.K. Rose. 2007. Depauperate small mammal communities in managed pine plantions in eastern Virginia. Journal of the Virginia Academy of Science In press.

Dueser, R.D., and H.H. Shugart, Jr. 1979. Niche pattern in a forest-floor small-mammal fauna. Ecology 60:108–118.

Dunaway, P.B. 1955. Late fall home ranges of three golden mice, *Peromyscus nuttalli*. Journal of Mammalogy 36:297–298.

Eads, J.H., and J.S. Brown. 1953. Studies on the golden mouse, *Peromyscus nuttalli aureolus* in Alabama. Journal of the Alabama Academy of Science 25:25–26.

Engstrom, M.D., and J.W. Bickham. 1982. Chromosome banding and phylo-genetics of the golden mouse, *Ochrotomys nuttalli*. Genetica 59:119–126.

Faust, B.F., M.H. Smith, and W.B. Wray. 1971. Distances moved by small mammals as an apparent function of grid size. Acta Theriologica 16:161–177.

Feldhamer, G.A., and K.A. Maycroft. 1992. Unequal capture response of sympatric golden mice and white-footed mice. American Midland Naturalist 128:407–410.

Feldhamer, G.A., and C.R. Paine. 1987. Distribution and relative abundance of the golden mouse (*Ochrotomys nuttalli*) in Illinois. Transactions of the Illinois Academy of Science 80:213–220.

Frank, P.A., and J.N. Layne. 1992. Nests and daytime refugia of cotton mice (*Peromyscus gossypinus*) and golden mice (*Ochrotomys nuttalli*) in south-central Florida. American Midland Naturalist 127:21–30.

Goodpaster, W.W., and D.F. Hoffmeister. 1954. Life history of the golden mouse, *Peromyscus nuttalli,* in Kentucky. Journal of Mammalogy 35:16–27.

Ivey, R.D. 1949. Life history notes on three mice from the Florida east coast. Journal of Mammalogy 30:157–162.

Kitchings, J.T., and D.J. Levy. 1981. Habitat patterns in a small mammal community. Journal of Mammalogy 62:814–820.

Lack, D.L. 1966. Population studies of birds. Oxford University Press, Oxford, United Kingdom.

Layne, J.N. 1960. The growth and development of young golden mice, *Ochrotomys nuttalli*. Quarterly Journal of the Florida Academy of Sciences 23:36–58.

Linzey, D.W. 1968. An ecological study of the golden mouse, *Ochrotomys nuttalli*, in the Great Smoky Mountains National Park. American Midland Naturalist 79:320–345.

Linzey, D.W., and A.V. Linzey. 1967a. Growth and development of the golden mouse, *Ochrotomys nuttalli nuttalli*. Journal of Mammalogy 48:445–458.

Linzey, D.W., and A.V. Linzey. 1967b. Maturational and seasonal molts of the golden mouse, *Ochrotomys nuttalli*. Journal of Mammalogy 48:236–241.

Loeb, S.C. 1999. Responses of small mammals to coarse woody debris in a southeastern pine forest. Journal of Mammalogy 80:460–471.

Lord, R.D., Jr. 1960. Litter size and latitude in North American mammals. American Midland Naturalist 64:488–499.

McCarley, W.H. 1958. Ecology, behavior and population dynamics of *Peromyscus nuttalli* in eastern Texas. Texas Journal of Science 10:147–171.

Mengak, M.T., and D.C. Guynn, Jr. 2003. Small mammal microhabitat use on young loblolly pine regeneration areas. Forest Ecology and Management 173:309–317.

Meserve, P.L. 1977. Three-dimensional home ranges of cricetid rodents. Journal of Mammalogy 58:549–558.

Miller, D.A., R.E. Thill, M.A. Melchoirs, T.B. Wigley, and P.A. Tappe. 2004. Small mammal communities of streamside management zones in intensively managed pine forests of Arkansas. Forest Ecology and Management 203:381–393.

Morzillo, A.T., G.A. Feldhamer, and M.C. Nicholson. 2003. Home range and nest use of the golden mouse (*Ochrotomys nuttalli*) in southern Illinois. Journal of Mammalogy 84:553–560.

Packer, W.C., and J.N. Layne. 1991. Foraging site preferences and relative arboreality of small rodents in Florida. American Midland Naturalist 125:187–194.

Patton, J.L., and T.C. Hsu. 1967. Chromosomes of the golden mouse, *Peromyscus nuttalli* Harlan. Journal of Mammalogy 48:637–639.

Pearson, P.G. 1953. A field study of *Peromyscus* populations in Gulf Hammock, Florida. Ecology 34:199–207.

Perry, R.W., and R.E. Thill. 2005. Small mammal responses to pine regeneration treatments in the Ouachita Mountains of Arkansas and Oklahoma, USA. Forest Ecology and Management 219:81–94.

Pruett, A.L., C.C. Christopher, and G.W. Barrett. 2002. Effects of a forested riparian peninsula on mean home range size of the golden mouse (*Ochrotomys nuttall*) and the white-footed mouse (*Peromyscus leucopus*). Georgia Journal of Science 60:201–208.

Redman, J.P., and J.A. Sealander. 1958. Home ranges of deer mice in southern Arkansas. Journal of Mammalogy 39:390–395.

Rose, R.K., and J.W. Walke. 1988. Seasonal use of nest boxes by *Peromyscus* and *Ochrotomys* in the Dismal Swamp of Virginia. American Midland Naturalist 120:258–267.

Schmid-Holmes, S., and L.C. Drickamer. 2001. Impact of forest patch characteristics on small mammal communities: A multivariate approach. Biological Conservation 99:293–305.

Seagle, S.W. 1985a. Patterns of small mammal microhabitat utilization in cedar glade and deciduous forest habitats. Journal of Mammalogy 66:22–35.

Seagle, S.W. 1985b. Competition and coexistence of small mammals in an east Tennessee pine plantation. American Midland Naturalist 114:272–282.

Shadowen, H.E. 1963. A live trap study of small mammals in Louisiana. Journal of Mammalogy 44:103–108.

Smith, M.H., R. Blessing, J.G. Chelton, J.B. Gentry, F.B. Golley, and J.T. McGinnis. 1971. Determining density for small mammal populations using grid and assessment lines. Acta Theriologica 16:105–125.

Smith, M.H., J.B. Gentry, and J. Pinder. 1974. Annual fluctuations in small mammal population in an eastern hardwood forest. Journal of Mammalogy 55:231–234.

Stickel, L.F. 1954. A comparison of certain methods of measuring home ranges of small mammals. Journal of Mammalogy 35:1–15.

Wallace, J.T. 1969. Some notes on the growth, development and distribution of *Ochrotomys nuttalli* (Harlan) in Kentucky. Transactions of the Kentucky Academy of Science 30:45–52.

Wilder, C.D., Jr., and R.D. Fisher. 1972. Occurrence of the golden mouse in southwestern Virginia. Chesapeake Science 13:326–327.

4
Community Ecology of the Golden Mouse

Cory C. Christopher and Guy N. Cameron

It is admitted that as a result of competition two similar species scarcely ever occupy similar niches, but displace each other in such a manner that each takes possession of certain peculiar kinds of food and modes of life in which it has an advantage over its competitor. . . . This once more confirms the thought mentioned earlier, that the intensity of competition is determined not by the systematic likeness, but by the similarity of the demands of the competitors upon the environment. (Gause 1934:19)

Interactions between species might be positive $(+,+)$, negative $(-,-)$, or mixed $(+,-)$, resulting in an array of possible types of interaction (e.g., competition, mutualism, predation, parasitism; Odum and Barrett 2005). These interactions might affect selection pressure on individuals, life-history parameters of populations, or species assemblages of communities.

Competition $(-,-)$ between different species can change the morphology of individual species (Dayan et al. 1989) or the structure of communities. Examples of the latter include the evolution of niche complementarity (McKinzie and Rolfe 1986, Ray and Sunquist 2001), spatial and temporal partitioning of resources (Kronfeld-Schor and Dayan 2003), and checkerboard geographic distributions (Gotelli and McCabe 2002). These effects allow coexistence of similar species within the same general geographic region and provide a means for maintenance of biodiversity in a system. Mutualism $(+,+)$ enhances growth and survival of individual species and also might alter community structure (Glynn 1983, Maser et al. 1978, Stachowicz 2001). Mutualism serves many functions, including increased access to food resources, reduced parasite load, and increased protection from predators.

Predation and parasitism are examples of mixed interactions $(+,-)$. Predation can act as an agent of natural selection or as a density-dependent factor regulating populations (Hanski et al. 2001). It can also affect species diversity and trophic structure by reducing interspecific competition among prey (Carpenter and Kitchell 1993, Power 1992, Power et al. 1996). Parasitism also can affect population survival (e.g., Burns et al. 2005, Fuller and Blaustein 1996) and community structure (e.g., Ostfeld and LoGiudice 2003, Schmitz and Nudds 1994).

Understanding the role of interspecific interactions in the biology and ecology of the golden mouse (*Ochrotomys nuttalli*) has been based primarily on descriptive studies with few empirical examinations used to validate hypotheses. Consideration of both types of study is important to identify possible interspecific interactions and set the stage for future studies on the ecological role of this species in communities.

Niche of the Golden Mouse

An understanding of the role of interspecific interactions in the biology of the golden mouse may begin with an assessment of those niche parameters that might be used by golden mice and other small mammals.

Distribution

The golden mouse occurs from southern Illinois, northern Kentucky, southern West Virginia, and eastern Virginia south to Mississippi, southern Louisiana, and central Florida, and west into eastern Texas and Oklahoma (see Chapter 2 of this volume for details). This distribution overlaps with several other species of small mammals whose niche characteristics are similar to those of the golden mouse, particularly the white-footed mouse (*Peromyscus leucopus*) and the cotton mouse (*Peromyscus gossypinus*).

Habitat and Microhabitat

Golden mice occur in moist, dense thickets, brushy areas, and thick woods, often in association with honeysuckle (*Lonicera* spp.), greenbrier (*Smilax* spp.), and other vines or in canebrakes (Linzey and Packard 1977). Specific habitat preferences depend on locality. Golden mice prefer dense deciduous woods with extensive, nearly impenetrable underbrush in Alabama (Eads and Brown 1953), upland and floodplain forest characterized by densely thicketed underbrush in the pine-oak region of east Texas, where cotton mice were nearly always found (McCarley 1958), and marginal thickets and mesic hammocks with dense shrubs such as saw palmetto (*Serenoa repens*), wax myrtle (*Morella cerifera*), yaupon (*Ilex vomitoria*), gallberry (*I. glabra*), sparkleberry (*Vaccinium arboretum*), arrowwood (*Viburnum dentatum*), blackhaw (*V. prunifolium*), and French mulberry (*Callicarpa americana*), as well as small trees with hanging arboreal vegetation, in Florida (Pearson 1954).

In the Great Smoky Mountains National Park, Tennessee, Linzey (1968) found golden mice in mixed Virginia pine-deciduous tree habitat [e.g., sumac (*Rhus* spp.), yellow poplar (*Liriodendron tulipifera*), sweet gum (*Liquidambar styraciflua*)], conifer-hardwood habitat [e.g., greenbrier, poison ivy (*Toxicodendron radicans*), and Japanese honeysuckle (*L. japonica*)], and habitat dominated by Japanese honeysuckle. Deer mice (*P. maniculatus*), cotton mice, white-footed mice, northern short-tailed shrews (*Blarina brevicauda*), jumping mice (*Napaeozapus insignis*), and smoky shrews (*Sorex fumeus*) also occur in these habitats.

In the Walker Branch Watershed, Tennessee, Dueser and Shugart (1978) reported *O. nuttalli*, *P. leucopus*, and the Eastern chipmunk (*Tamias striatus*) from four habitats: oak-hickory (*Quercus* spp. and *Carya* spp.), chestnut oak (*Q. prinus*), pine (*Pinus* spp., mostly shortleaf pine, *P. echinata*), and yellow poplar. Kitchings and Levy (1981) found the same three species of small mammals in the Oak Ridge National Environmental Research Park from habitats with the Eastern red cedar (*Juniperus virginiana*), shortleaf pine, and chestnut oak. In this same area, Seagle (1985) found *O. nuttalli* only in habitats with dense growth of vines, although *P. leucopus* also was associated with understory thickness.

Southern Illinois and northern Kentucky lie at the northern edge of the range of golden mice (see Chapter 2 of this volume for details). In southern Illinois, they occupy mesic wooded habitats with dense undergrowth of honeysuckle, greenbrier, poison ivy, cane (*Arundinaria gigantea*), brier (*Schrankia* spp.), and rhododendron (*Rhododendron* spp.; Andrews 1963, Blus 1966, Layne 1958). In northern Kentucky, golden mice occupy pine-hardwood habitats with heavy density of spice bush (*Lindera benzoin*), dogwood (*Cornus* spp.), greenbrier, blackberry (*Rubus* spp.), honeysuckle, and wild grape (*Vitis* spp.; Goodpaster and Hoffmeister 1954).

The golden mouse has been considered a habitat specialist, preferring deciduous and mixed deciduous-evergreen forest in Tennessee (Dueser and Hallett 1980). However, studies in southern Illinois determined that golden mice used a variety of habitats (Andrews 1963, Blus 1966, Morzillo et al. 2003), indicating a more generalist strategy. High diet diversity in these animals supported this conclusion (Knuth and Barrett 1984; see *Food* subsection). Differences in habitat specificity of golden mice between localities could reflect geographic patterns or differences in the structure of small mammal communities in these habitats. Differences in habitat use might affect interspecific interactions, as competition for resources might be greater where golden mice exhibit a more generalist strategy.

Food

Golden mice have been characterized as granivores (Goodpaster and Hoffmeister 1954). Goodpaster and Hoffmeister (1954) identified seeds from golden mouse feeding platforms in northern Kentucky as follows: most abundant, sumac, *Rhus* spp.; wild cherry, *Prunus avium*; dogwood, and greenbrier; average abundance, oak, *Quercus* spp.; bindweed, *Polygonum* spp.; peppervine, *Ampelopsis arborea*; pokeweed, *Phytolacca americana*; tick clover, *Desmodium* spp.; and least numerous, bittersweet, *Celastrus scandens*; brome grass, *Bromus* spp.; locust, *Robinia* spp.; bed straw, *Galium* spp.; clover, *Trifolium* spp.; milkweed, *Asclepias* spp.; corn, *Zea mays*; wild bean, *Phaseolus polyslachoys*; hog-peanut, *Amphicarpa bracteata;* scurf pea, *Psoralea esculenta*; box elder, *Acer negundo*; basswood, *Tilia americana*; and blackhaw.

Blus (1966) recovered food items from 28 nests in southern Illinois. In rank order, seeds of oak, poison ivy, bedstraw, and blackberry accounted for 54.2 percent of the food items. Grape, sassafras (*Sassafras albidum*), and climbing false buck-wheat (*Polygonum scandens*) accounted for 20.9 percent of seeds. Seeds from plants that were important indicators of golden mouse habitat such as Japanese honeysuckle, giant cane, and catbrier (*Smilax* spp.) were not found—suggesting that they were unimportant in the diet. deRageot (1964) found that greenbrier, sumac, dogwood, and wild cherry seeds were preferred by golden mice in the Great Dismal Swamp of Virginia.

Linzey (1968) found seeds in 73 percent of 44 nests analyzed in Tennessee, including those of wild cherry (49 percent of food items), dogwood (38 percent), greenbrier (19 percent), yellow poplar (14 percent), and oak (8 percent). Greenbrier was the dominant food item identified from stomach contents, as was blackberry from one study area. These items were uncommon in nests, suggesting that mice ate entire seeds while foraging. There were insects in 47.6–57.0 percent of stomachs. In this same area, *P. gossypinus* ate 68 percent animal matter—mostly Coleoptera, Lepidoptera, and Araneida (Calhoun 1941).

Barrett and co-researchers used laboratory-feeding trials to determine food preferences of golden mice and white-footed mice from different geographic locales. Knuth and Barrett (1984) offered native food plants characteristic of habitats in Kentucky and South Carolina from which golden mice were obtained and from habitats in Ohio from which white-footed mice were obtained. Fruits of sumac, blackberry, honeysuckle, cherry, and cracked corn were offered. Golden mice preferred blackberry fruit, whereas white-footed mice preferred cherry fruit. Diet diversity and evenness was higher for golden mice than white-footed mice, indicating that golden mice were more of a generalist species than had been concluded from habitat studies. In this study, sumac was not selected by either species. In another laboratory-feeding study, Jewell et al. (1991) demonstrated that golden mice fed on an all-sumac diet tended to lose weight, which they attributed to either reduced ingestion rates, toxicity of sumac fruit, or appetite depression.

Peles et al. (1995) offered golden mice fruits of the Eastern red cedar and Japanese honeysuckle because these plant species dominated a canyon habitat in Madison County, Kentucky. Mice fed cedar fruits had lower rates of ingestion, lost more body mass, and had lower survival than mice fed honeysuckle fruits. Fruits of honeysuckle and juniper were equal in caloric content. Crude protein content of honeysuckle fruit was higher (9.75 percent; juniper 8.41 percent), but honeysuckle fruit contained more phenols (2.20 percent; juniper 1.23 percent). They concluded that food quality affected diet item selection.

O'Malley et al. (2003) tested whether food quality affected diet selection of *O. nuttalli* and *P. leucopus* from a forested bottomland in Georgia. They offered acorns of water oak (*Quercus nigra*) and white oak (*Q. alba*) and berries from Chinese privet (*Ligustrum sinense*) because these plants were dominant in this habitat. White-footed mice preferred water oak acorns (78 percent), followed by white oak acorns (15 percent) and Chinese privet berries (7 percent). Golden mice primarily ate water oak acorns (91 percent) and few privet berries (9 percent). Selection of food items reflected higher tannin content in white oak (13.3 percent versus 2.3 percent for water oak).

Nests

Chapter 9 of this volume discusses nesting behavior of golden mice in detail. Briefly, golden mice have three characteristic arboreal structures throughout their range: a nest for shelter, a globular nest for rearing young (Barbour 1942), and feeding platforms used for depositing, opening, and consuming seeds (Blus 1966, Goodpaster and Hoffmeister 1954). The globular nests used as a home site consist of leaves, shredded bark, and grass (Blus 1966, Frank and Layne 1992, Goodpaster and Hoffmeister 1954, Ivey 1949, Linzey 1968, Morzillo et al. 2003). Golden mice also use ground nests (Easterla 1968, Ivey 1949, Morzillo et al. 2003), and in some habitats they use only ground nests (east Texas, McCarley 1958; southern Illinois, Andrews 1963; and Alabama, Eads and Brown 1953). Exclusive use of ground nests might occur only in certain types of habitat, such as pine-oak and pine forest in Texas (McCarley 1958), eroded areas with generally sparse undercover in southern Illinois (Andrews 1963), and swamp habitat in Alabama (Eads and Brown 1953). Pearson (1954), however, reported only two arboreal nests from marginal thickets and mesic hammocks in northern Florida, where there was dense vegetative cover, including maple and oak trees and a variety of vines and dense shrubs. He also found use of burrows under rotting logs and subterranean chambers under an oak tree. Morzillo et al. (2003) found that females from southern Illinois used more arboreal nests, whereas males used more ground nests; only males made exclusive use of ground nests.

Cotton mice and white-footed mice use arboreal nests (Dooley and Dueser 1996) and could compete for nest sites with golden mice. Both *P. gossypinus* and *O. nuttalli* use ground nests in hammocks near the coast in northern Florida (Pearson 1954). In southcentral Florida, only 2 of 77 refugia of *O. nuttalli* were aboveground in shrubs (Frank and Layne 1992). On the other hand, Ivey (1949) reported that *P. gossypinus* nested on the ground on a barrier island off the eastern coast of Florida where golden mice occupied arboreal nests.

Temporal Use of Space

The activity of golden mice is crepuscular and nocturnal (Kennedy et al. 1973), with days spent primarily in refugia. *O. nuttalli* and *P. gossypinus* had greater refuge site fidelity during winter than summer in a ridge sandhill habitat in Florida (Frank and Layne 1992). In winter, they used fewer refugia (*O. nuttalli*: 3.5 refugia in winter and 5.3 in summer; *P. gossypinus*: 4.0 refugia in winter and 6.2 refugia in summer), spent more days per refuge (*O. nuttalli*: 6.2 days per refuge in winter and 4.6 days per refuge in summer; *P gossypinus*: 6.2 days per refuge in winter and 4.0 days per refuge in summer), and switched refuges less often (*O. nuttalli*: 4.7 switches in winter and 11.6 switches in summer; *P. gossypinus*: 5.7 switches in winter and 12.0 switches in summer).

Golden mice are arboreal animals (Blus 1966, Goodpaster and Hoffmeister 1954, Morzillo et al. 2003). Christopher and Barrett (2006) set traps on the ground and in vegetation at 1.5 m and 4.5 m above the ground. Both *O. nuttalli* and *P. leucopus* occurred at all heights (as discussed in the *Empirical Studies of Interspecific Interactions* subsection).

Golden Mice and Interspecific Interactions

Interspecific Competition

The presence or absence of interspecific competition between golden mice and other small mammals has been inferred by descriptions of habitat associations or patterns of trapping. Although such studies might be suggestive, they are insufficient to make firm statements or predictions about interspecific interactions, in part because it is difficult to tease apart competition from differential susceptibility to predators (i.e., apparent competition) or from preferences for different microhabitats.

Descriptive Evidence for Interspecific Competition

In east Texas, McCarley (1958) determined that golden mice, cotton mice, and white-footed mice had similar seasonal fluctuations in population density and breeding and were likely affected by similar environmental factors. He nearly always found golden mice along with cotton mice in a floodplain forest habitat, although golden mice were less abundant. The ratio of golden mice to cotton mice was about 1:2 in the xeric margin between floodplain and upland forest, and golden mice were frequently the only species in upland forest (McCarley 1958). The distribution of golden mice appeared to be related to understory density. Co-occurrence of these species in the same floodplain forest, however, does not necessarily indicate that they interacted. The shift toward dominance by *P. gossypinus* in the forest margin, though, could reflect interspecific interactions, particularly given that both species nested on the ground—no arboreal nests for *O. nuttalli* were evident. Alternatively, the forest margin could be a poorer habitat for *O. nuttalli* than the floodplain or upland forest, but this possibility was not explored.

In the Great Smoky Mountains National Park, Tennessee, Linzey (1968) found spatial isolation between *O. nuttalli* and three species of *Peromyscus* in a study area dominated by the Virginia pine (*P. virginiana*), sumac, and yellow poplar. *O. nuttalli* occupied the central portion of the study area and *P. leucopus*, *P. maniculatus*, and *P. gossypinus* were restricted to habitat edges and stone walls that occurred throughout the study area.

In 16 counties in southern Illinois, Feldhamer and Paine (1987) livetrapped golden mice, but not *P. leucopus*, more often than expected in aboveground vegetation in suitable habitat. Although this result indicated spatial separation between these species, there was not a significant correlation between their relative abundances. However, Feldhamer and Maycroft (1992) did find a significant negative relationship between number of golden mice and number of white-footed mice on 21 sites in shortleaf pine (*P. echinata*) and loblolly pine (*P. taeda*) in southern Illinois. They also found that the number of trap stations that captured both species was significantly less than expected. These results suggested a negative interaction between these two rodent species. To determine a possible mechanism for these interspecific interactions, Feldhamer and Maycroft (1992) determined that the mean number of captures and recaptures, the mean number of individuals captured only once, the mean number of different traps in which an individual was captured, and the mean number of different traps in which an individual was captured for mice taken more than one time were all significantly less for golden mice. They argued that these results showed that spatial segregation between golden mice and white-footed mice resulted from behavioral exclusion of golden mice by white-footed mice.

Faust et al. (1971), however, did not find a difference between the number of *O. nuttalli* and the number of *P. gossypinus* captured only once in a lowland mesic-hardwood forest in South Carolina. They also found that home range and average distance moved between captures was not larger for *P. gossypinus*, as predicted for a competitively dominant species (Calhoun 1941). Variation in geography and microhabitat between Illinois and South Carolina that affected interspecific interactions might explain these different results.

Pearson (1954) used live trapping to study populations of cotton mice and golden mice in a mesic hammock in Florida. Density of cotton mice peaked from late September through early December. As the density of cotton mice decreased in mid-December, the density of golden mice increased, but never reached that of cotton mice. This reciprocal relationship also was evident in trappability of these species. Golden mice were captured only a few times during the peak density of cotton mice, but their trappability increased when the density of cotton mice decreased. The average number of trap nights necessary to capture one golden mouse during the peak abundance of cotton mice from mid-September to early December was 77.5, whereas no golden mice were captured earlier. When the density of cotton mice decreased, the average number of trap nights necessary to capture one golden mouse declined to 8.9 (mid-December to mid-March; Pearson 1954).

Pearson (1954) considered four hypotheses to explain this increase in captures of *O. nuttalli* with the decrease of *P. gossypinus*: (1) increase in total number of *O. nuttalli*, (2) probability of capture of *O. nuttalli* increased, (3) individual *O. nuttalli* moved in from another habitat, and (4) individual *O. nuttalli* moved from arboreal to ground habitat. Pearson's data did not substantiate hypothesis 1 or 2, and he rejected hypothesis 4 because golden mice used only two arboreal nests. He concluded, therefore, that golden mice moved onto the trapping grid when interspecific pressure from *P. gossypinus* was reduced, a result suggesting that *P. gossypinus* was competitively dominant. This conclusion agreed with the findings of Feldhamer and Maycroft (1992) but differed from conclusions of Dueser and colleagues discussed later in this subsection.

Packer and Layne (1991) studied golden mice in the southern ridge sandhill and sand pine scrub habitats in Florida. Sandhill grids contained a widely spaced overstory of slash pine (*P. elliotti*), whereas the scrub grid consisted of dense sand pines (*P. clausa*). Dense shrub layers of oaks (*Q. chapmanii, Q. myrtifolia, Q. laevis, Q. germinata*), scrub hickory (*C. floridana*), and palmetto (*Serenoa repens, Sabal etonia*) occurred in both habitats. The sandhill habitat had openings of various sizes consisting of exposed sand. Small mammals in these habitats included the Florida mouse (*Podomys floridanus*), old-field mouse (*P. polionotus*), cotton mouse, golden mouse, and hispid cotton rat (*Sigmodon hispidus*). These species varied in morphology, habitat preference, and foraging behavior, leading to the prediction that they would occupy different microhabitats. Packer and Layne found that smaller old-field mice, cotton mice, and golden mice located bait more efficiently than did larger *Podomys* or *Sigmodon*. Golden mice and cotton mice were more arboreal than the other species; *P. gossypinus* used the arboreal habitat more frequently and golden mice less frequently than suggested by their morphology and habits. These results and the relatively broad overlap in foraging niches among these species led Packer and Layne (1991) to conclude that competition was not excluding any species of small mammal and was not as important in structuring this small mammal community.

In a subsequent study in the southern ridge sandhill habitat, Frank and Layne (1992) investigated nesting and daytime refugia of *P. gossypinus* and *O. nuttalli*. They located mice in day refugia with radiotelemetry during summer and winter. Cotton mice used ground holes, tree cavities, and especially tortoise burrows as day refuges during both summer and winter. They suggested that use of tortoise burrows might be related to their stable microclimate. These burrows afford protection from high summer temperature and periodic wildfires characteristic of this sandhill habitat. Alternatively, use of tortoise burrows might increase risk of predation—five instances of weasel predation were noted. Golden mice used ground day refugia almost exclusively during both seasons. They were in soil depressions under layers of pine needles, oak and hickory leaves, and twigs. The authors thought that ground nests offered more protection from predators than the more exposed arboreal nests because they had escape tunnels.

Frank and Layne (1992) also noted that nest material differed between the two species: *P. gossypinus* used saw palmetto fibers, lichens, and cotton, whereas *O. nuttalli* used a variety of materials, including saw palmetto fibers, Spanish moss, leaves, twigs, grasses, and cotton. They concluded that divergence in refuge selection between *P. gossypinus* and *O. nuttalli* might allow these species to coexist. Mechanisms maintaining this different pattern of habitat use were unclear, but no evidence for competition for refuge sites was found. Similarly, Goodpaster and Hoffmeister (1954) concluded that it was unlikely that there was competition for food or nest sites between golden mice and white-footed mice in northern Kentucky because their nest sites were spatially separate, albeit in the same habitat.

Dueser and Shugart (1978, 1979) studied the small mammal community consisting of *O. nuttalli, P. leucopus, Blarina brevicauda*, and *Tamias striatus* in the Walker Branch Watershed, Tennessee. They livetrapped in four forest stands dominated by oak-hickory (*Quercus* spp. and *Carya* spp.), chestnut oak (*Q. prinus*), pine (mostly *P. echinata*), and yellow poplar. Captures were 66 percent *P. leucopus*, 27 percent *Tamias*, 5 percent *Ochrotomys*, and 2 percent *Blarina*. Although most captures were at sites where no other species was trapped, overlap in use of trap sites did occur between *Blarina* and *Peromyscus*, *Blarina* and *Ochrotomys*, and *Peromyscus* and *Ochrotomys*. Only *Peromyscus* and *Tamias* showed significant association.

Dueser and Shugart (1978) analyzed 29 vegetation attributes using discriminate function analyses to characterize microhabitats of each small mammal species. The authors noted that comparisons between species for 20 of these variables were "considerably" greater than what would be expected from sample variation (Table 4.1). Golden mouse microhabitat differed conspicuously from the average small mammal capture site by having a pronounced shrub component. The two species most dissimilar in body size and general ecology, *P. leucopus* and *T. striatus*, had the most similar qualitative microhabitat characteristics. The species most similar in appearance, body size, and general ecology, *P. leucopus* and *O. nuttalli*, were the least similar qualitatively in microhabitat characteristics. These results were consistent with relatively minor interspecific interactions between *P. leucopus* and *T. striatus* but with more intense interspecific interactions between *P. leucopus* and *O. nuttalli*.

TABLE 4.1. Habitat variables that distinguish trap sites at which white-footed mice (*P. leucopus*), Eastern chipmunks (*T. striatus*), and golden mice (*O. nuttalli*) were captured.

Variable	P. leucopus	T. striatus	O. nuttalli
Thickness of woody vegetation	*		****
Shrub cover		**	****
Overstory tree dispersion	*		
Understory tree dispersion	****	****	
Woody stem density		*	****
Short woody stem density			****
Woody foliage profile density		*	****
Number of woody species	***	**	
Herbaceous stem density		*	****
Short herbaceous stem density		*	****
Herbaceous foliage profile density		*	****
Number of herbaceous species			***
Evergreenness of overstory	****		****
Evergreenness of shrubs	****	****	
Evergreenness of herb stratum			**
Tree stump density			****
Tree stump size	****		****
Litter-soil depth		**	
Litter-soil density			*
Litter-soil compactibility	**		**

Source: After Dueser and Shugart (1978).

Notes: Sites where each species was captured were compared to the pooled sites where individuals of the other species were captured by an analysis of variance of each habitat variable.

$* \leq 0.05$; $** \leq 0.025$; $*** \leq 0.01$; $**** \leq 0.001$.

Dueser and Shugart (1978) concluded that the more abundant *P. leucopus* was a microhabitat generalist associated with common patch types, and the less common *O. nuttalli* was a habitat specialist restricted to a rare patch. They suggested that such habitat separation would be expected in an ecologically saturated community, albeit this community had relatively low small mammal species diversity.

Dueser and Shugart (1979) used multiple discriminate function analysis to analyze realized niche patterns of these same small mammal species. Each species was segregated on at least one discriminant axis. The golden mouse was a midsuccessional species favoring evergreen forest with heavy undergrowth. White-footed mice and Eastern chipmunks responded to deciduous forest with a dense shrub layer. The greatest niche overlap was between *Peromyscus* and *Tamias* and the least overlap was between *Peromyscus* and *Ochrotomys*. The niche breadth of *Ochrotomys* was narrowest, that of *Tamias* was intermediate, and that of *Peromyscus* was widest. Dueser and Shugart (1979) used niche position, niche breadth, and population abundance to compute a community niche pattern. *O. nuttalli* occupied an extreme niche position, had little variability in niche configuration, and occurred in low abundance even in their optimal habitat, characteristics of a habitat specialist poorly adapted to these habitats. *T. striatus* was intermediate between *P. leucopus* and *O. nuttalli*, with a niche less variable than *P. leucopus* but less extreme than golden mice. These small mammals were separated completely along three structural niche dimensions and species positions appeared to exhibit niche complementarity (Schoener 1974). Hence, Dueser and Shugart (1979) concluded that these data indicated competitive coexistence and noted that experimental studies would be necessary to validate this conclusion.

Dueser and Hallett (1980) used these trapping data in a multiple regression analysis to quantify competitive effects. All elements in a community matrix were negative, indicating intense competition among species in spring, summer, and winter. White-footed mice, Eastern chipmunks, and golden mice exhibited increasing competitive ability and increasing habitat selectivity. They concluded that the high competitive ability of the relatively rare golden mouse might explain how it persisted in a community with more abundant species and how it recovered from a low population density. They suggested that *O. nuttalli*, as a superior competitor and habitat specialist, might occupy an "included niche" relative to other species in the community. They also suggested that removal experiments would be necessary to validate these conclusions. For example, removal of *Tamias* or *Peromyscus* should have no effect on the habitat specialist *Ochrotomys*; removal of *Peromyscus* should have no effect on the competitively superior *Tamias*; removal of the competitively superior *Ochrotomys* should drive niche shifts by *Tamias* and *Peromyscus*; and removal of *Tamias* should elicit a niche shift by *Peromyscus*. They cautioned that niche shifts in *Tamias* and *Peromyscus* might be difficult to detect because these species are habitat generalists.

Kitchings and Levy (1981) selected another habitat in the same watershed in which Dueser and colleagues had worked to determine whether they would obtain the same microhabitat preferences for *P. leucopus, O. nuttalli, T. striatus*, and

B. brevicauda. The habitat they selected had an overstory dominated by the Eastern red cedar, shortleaf pine, and chestnut oak. Using similar measures of microhabitat and methods of data analysis as Dueser and colleagues, Kitchings and Levy (1981) found that species generally followed the same pattern of habitat use as previously reported. Results might have differed because Dueser and colleagues worked in a late successional plant community, whereas Kitchings and Levy conducted their studies in early successional stages. During autumn, *P. leucopus* occurred in open, swampy areas with more grasses and shrubs and less compacted soil; during summer, this species was in wooded, rocky sites. Golden mice were livetrapped in open areas and along swamp edges during summer, but they were in cedar and heavily wooded areas during autumn. Chipmunks were a woodland generalist, using various microhabitats within forests. Although they did not do a niche analysis, Kitchings and Levy concluded that habitat separation would be an important part of the niche because of its association with resources such as food supply and nesting sites.

Contrary to the conclusion of Dueser and colleagues, Knuth and Barrett (1984) concluded that golden mice were more of a generalist because they had a more even apportionment among food items than *P. leucopus* in laboratory-feeding trials. They attributed this difference to larger cheek pouches in *O. nuttalli* that afforded an opportunity to gather multiple types of seed and to return to a feeding platform or nest to eat. They also surmised that *O. nuttalli* might move more between feeding bouts than *P. leucopus*, thereby affording them access to a wider diversity of seed types. This behavior, as well as increased energy efficiency (e.g., higher assimilation rates), would confer an advantage to *O. nuttalli* over *P. leucopus*.

Morzillo et al. (2003) used radiotelemetry to track *O. nuttalli* in a habitat dominated by persimmon (*Diospyros virginiana*), honey locust (*Gleditsia triacanthos*), black locust (*Robinia pseudocarcia*), the Eastern red cedar, autumn olive (*Elaeagnus umbellata*), oaks, hickories, poison ivy, and grape in southern Illinois. Portions of their study area consisted of fragmented patches of woods with dense undergrowth. Mice used various portions of the habitat, ventured into grassy fields with sparse cedars, and crossed roads, which suggested that golden mice might be more of a habitat generalist than other authors had suggested.

In summary, studies in Tennessee concluded that golden mice were habitat specialists and superior competitors. Studies from southern Illinois and Kentucky, however, used habitat occupancy and diet items to determine that golden mice were habitat generalists. Such conflicting perspectives could indicate geographic or site-specific differences between golden mice and other small mammal species in their respective communities (see Chapter 3 of this volume for additional information regarding this hypothesis).

Results from descriptive studies focusing on abundance differ on whether competitive interactions exist between golden mice and other small mammal species. For example, Pearson (1954) and McCarley (1958) found a negative relation between abundance of *O. nuttalli* and *P. leucopus*. Feldhamer and Paine (1987) detected a vertical separation in space use between these two species, perhaps indicating competition. Dueser and colleagues concluded that intense competition between golden mice and other small mammals in Tennessee yielded a situation of competitive coexistence, with golden mice being a superior competitor. On the other hand, Feldhamer and Maycroft (1992) also discovered a spatial separation between golden mice and *P. leucopus,* but they came to an opposite conclusion, namely that golden mice were excluded by *P. leucopus.* This difference in the role of competition in small mammal communities containing golden mice could reflect geographic or site-specific microhabitat differences. The presence of such differences was reinforced in studies by Packer and Layne (1991) and Frank and Layne (1992) in Florida. They concluded that interspecific competition likely was not a factor in community structure because broad microhabitat overlap did not exclude any species and selection of refuge sites diverged between golden mice and white-footed mice. Goodpaster and Hoffmeister (1954) reached the same conclusion after finding separation of nest sites between *O. nuttalli* and *P. leucopus* in northern Kentucky.

Empirical Studies of Interspecific Competition

Few investigators have conducted experiments or collected experimental data to determine whether niche expansion results when a potential competitor is removed. Such studies could help clarify the conflicting results of the descriptive studies discussed earlier.

Seagle (1985) studied *B. brevicauda, P. leucopus,* and *O. nuttalli* on two live trapping grids in a loblolly pine plantation on the Oak Ridge National Environmental Research Park, Tennessee. Microhabitat variables described by Dueser and Shugart (1978) were measured at each capture site and at 30 randomly identified trap sites (of 64 trap sites in each grid). Both grids were operated for 5 weeks and then *P. leucopus* were removed from one grid. Seagle relied on early studies indicating that *P. leucopus* was a generalist species to infer that their removal would allow niche expansion by the more specialized golden mouse. Trapping continued for 7 weeks on an experimental (*P. leucopus* removed) and a control (nonremoval) grid.

The density of golden mice more than tripled on the experimental grid because of immigration of unmarked mice from surrounding habitat, but the density of *Blarina* remained similar on the control and experimental grids. Discriminant function analysis and 95 percent confidence ellipses showed that microhabitat affinities for all three species were separated on the control grid: *Blarina* inhabited areas with moderate fallen log abundance and deep litter-soil profile, *P. leucopus* were in areas with abundant fallen logs and dense understory, and golden mice occupied open areas with little ground structure (Figure 4.1A). Golden mice shifted microhabitat occupancy on the experimental grid to use more fallen logs, dense canopy structure, and heavy understory cover (Figure 4.1B). *Blarina* also shifted microhabitat use, but this shift was not significantly different from the control grid.

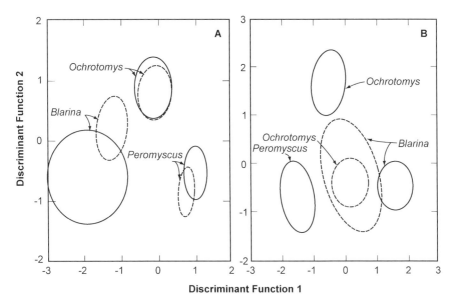

FIGURE 4.1. Microhabitat use by *Blarina*, *Ochrotomys*, and *Peromyscus* displayed as 95 percent confidence ellipses around species centroids for (a) control and (b) experimental (removal) grids. *P. leucopus* was removed from the experimental grid during the sixth week of trapping. Solid ellipses represent data taken before *P. leucopus* removal, and dashed ellipses represent data taken after removal. Discriminant Function 1 (DF1) was positively correlated with litter-soil depth, woody stem density, fallen log size, and fallen log dispersion and negatively correlated with fallen log abundance and climbing vine density. Discriminant Function 2 (DF2) was negatively correlated with woody stem density and positively correlated with overstory tree dispersion, understory tree dispersion, and fallen log dispersion. From Seagle (1985) with permission.

    Seagle (1985) computed niche position (distance from center of the discriminant space to a species centroid; a measure of the likelihood a species was in favorable microhabitat; Shugart and Patten 1972), niche breadth (coefficient of variation of the distance from a species centroid to individual observations), and niche overlap (percentage of 95 percent ellipse for a species overlapped by the ellipse of another species). Both golden mice and *Blarina* shifted niche position toward more favorable microhabitats with *P. leucopus* removed. Additionally, the niche breadth of both golden mice and *Blarina* increased, complementing a concomitant decrease in niche position. There was strong interspecific interaction between white-footed mice and golden mice pairs because niche overlap between these pairs showed little variation between trap periods. This ecological release demonstrated interspecific competition between *P. leucopus* and *O. nuttalli*. Because the response was rapid and new golden mice quickly moved onto the experimental grid, the likely mechanism was interference competition. Hence, the generalist *P. leucopus* was competitively dominant to the specialist *O. nuttalli*.

Corgiat (1996) repeated this experiment in southern Illinois. Golden mice did not exhibit a shift in microhabitat use on a grid with white-footed mice removed compared to a grid with white-footed mice present. Similarly, measures of niche breadth, niche overlap, and niche position differed among seasons, indicating the lack of a clear direction of interspecific competition.

Christopher and Barrett (2006) used a replicated design to study the interaction between *O. nuttalli* and *P. leucopus* in an upland deciduous forest dominated by water oak and American beech (*Fagus grandifolia*) and a bottomland deciduous forest dominated by river birch (*Betula nigra*) and water oak in Georgia. Both sites also contained Chinese privet, greenbrier, Amur honeysuckle (*Lonicera maackii*), and Japanese honeysuckle. They established 8 live trap grids each with 12 trap stations; a Sherman live trap was placed on the ground and another 1.5 m high on the trunk of a tree at each trap station. Four trap sites in each grid contained a trap placed 4.5 m in the canopy. They removed *P. leucopus* from four grids to assess the effects on golden mouse demography or use of habitat space. After removal of white-footed mice, there was no change in the relative abundance of golden mice compared to control grids, as previously demonstrated by Seagle (1985). However, golden mice changed use of the three-dimensional habitat after removal of white-footed mice; they were captured more frequently in ground traps than before removal. The authors suggested that abundant food resources, such as mast crop of water oak acorns, and heavy vegetative cover of Chinese privet dampened competition between these two small mammal species.

In summary, empirical results indicated competitive inhibition of golden mice by *P. leucopus*. Seagle (1985) found both an increase in density and the use of a wider array of microsites by golden mice after *P. leucopus* was removed. Christopher and Barrett (2006) also reported niche expansion (i.e., increased use of habitat space) by golden mice when *P. leucopus* was removed but not an effect on population abundance of golden mice. Contrary to the conclusion by Dueser and colleagues that golden mice, a specialist, were competitively dominant, these studies found that *P. leucopus* constrained the niche of golden mice likely by interference competition. A similar conclusion was reached by Feldhamer and Maycroft (1992) in southern Illinois. Thus, empirical studies do not verify the prediction of Dueser and Hallett (1980) that removal of *P. leucopus* would have no effect on *O. nuttalli*, although Corgiat (1996) did not find an effect. These data also might indicate that generalist or specialist tendencies of golden mice are highly influenced by geography or microhabitat characteristics. Nevertheless, some caution should be exercised when making such extrapolations from experimental studies. Studies by Seagle (1985) and Corgiat (1996) used only one removal and one control grid and it would be difficult to tease out site- or habitat-specific responses of golden and white-footed mice from effects of a lack of replication of study grids (Hurlbert 1984).

Predation

There has been little quantification of the extent to which golden mice are taken as prey, which predators might be important, or how predators might affect their populations. Maehr and Brady (1986) reported three golden mice eaten by bobcats (*Lynx rufus*) in Florida based on stomach analyses. Frank and Layne (1992) noted that five radiocollared *P. gossypinus* in Florida were partially eaten by weasels, but they did not record similar predation on golden mice. They suggested that ground nests well concealed beneath leaf litter offer more protection from predators than arboreal nests. They also observed tunnels from ground nests could be used to escape from predators (see also Easterla 1968, Packard and Garner 1964) and felt that ground nests would require less time to construct than arboreal nests, thereby decreasing exposure to predators. Others, however, suggested that arboreal nests with associated tangles of thick vegetation reduced the risk of predation compared to ground nests (Klein and Layne 1978, Morzillo et al. 2003, Wagner et al. 2000). No empirical studies have been conducted to test which aspects, if any, of golden mouse nest construction and placement offer protection from predators. These ideas and predator avoidance behavior by golden mice warrant further study.

Parasitism

Golden mice host a variety of internal and external parasites, including a cestode (*Taenia rileyi*), nematodes, mites, fleas, lice, ticks, and botflies (Goodpaster and Hoffmeister 1954, Linzey 1968, Pearson 1954). In Chapter 10 of this volume, Durden summarizes ectoparasites and the role of golden mice in vector-borne diseases. The effects of parasites on the ecology of the golden mouse need further investigation. It has been observed, however, that golden mice host botflies (*Cuterebra*) less often and less frequently than other small mammal species (see Durden of this volume). Here, we focus on this host–parasite relationship for *O. nuttalli* and *P. leucopus*.

Dunaway et al. (1967) reported botfly infestations in golden mice and white-footed mice from the Atomic Energy Commission Reservation near Oak Ridge, Tennessee; this was the first report of botflies in golden mice. The infestation rate in *P. leucopus* ranged from 19.1 percent to 33.3 percent from June to December over a 7-year period. Three parasitized *O. nuttalli* were captured in August; two of these individuals had two botflies and the other had one botfly and one botfly scar. All botflies were in the inguinal region. Dunaway et al. (1967) did not find any effect of botfly parasitism on body mass, testes, or activity, although they did find that the hematocrit percentage and hemoglobin concentration were lower and the total leukocyte number was greater in infected *P. leucopus*.

Jennison et al. (2006) recorded botflies (*C. fontinella*) in golden mice and white-footed mice in a bottomland forest near Athens, Georgia. Parasitism by botflies in white-footed mice peaked from early June to mid-August (37.1 percent infected in 2001, 20.7 percent in 2004), with a second peak from mid-September to November (41.7 percent). Botfly parasitism of golden mice occurred from mid-June to mid-August (23.3 percent infected in 2001; 3.6 percent in 2004), with late season parasitism from late August to mid-October (12.5 percent in 2004). White-footed mice were parasitized more often and more frequently than golden mice. These investigators attributed higher rates of parasitism in white-footed mice to greater levels of activity and greater three-dimensional home range size. Pruett et al. (2002) reported that home range size did not differ for these species in the same habitat in Georgia based on an analysis of two-dimensional home range sizes. Future studies addressing the activity pattern of these two small mammal species based on a three-dimensional perspective are needed.

Social Interactions

Several investigators have analyzed use of habitat space and home range size of golden mice (see Chapter 3 of this volume for details). Pruett et al. (2002) used live trapping and radiotelemetry to quantify home range size of golden mice and white-footed mice in a bottomland forest in Georgia. Home range sizes did not differ significantly between these small mammal species, but they proposed that analysis of home range size should include the vertical dimension.

Christopher and Barrett (2007) analyzed data on double captures taken during 2001 and 2004 in a bottomland forest in Georgia. They found 14 instances in which white-footed mice and golden mice were captured simultaneously in the same trap, 79 instances of two white-footed mice in the same trap, and 10 instances of two golden mice in the same trap. Although the proportion of interspecific double captures was small, this result might indicate tolerance between these species, given that the individuals captured together were not wounded. In view of the empirical evidence showing changes in use of habitat space by golden mice upon removal of *P. leucopus*, these double-capture data suggest that the mechanism of competition might be exploitative rather than interference. This view, however, differs from conclusions of interference competition between golden mice and white-footed mice reached from results of studies in southern Illinois (Feldhamer and Maycroft 1992), Tennessee (Seagle 1985), and Georgia (Christopher and Barrett 2006). Despite the fact that golden mice are often referred to as intraspecifically docile or social, there is little empirical evidence to extend such sociality to interspecific interactions. Further studies, such as pairwise aggression trials in the field, are necessary to elucidate the nature of any behavioral interactions between golden mice and white-footed mice. Although there are no empirical data supporting any interspecific positive relationships, Christopher and Pruett (personal communication) used radiotelemetry to observe a golden mouse and a white-footed mouse using the same nest, although they could not confirm simultaneous occupancy. Social interactions are a fertile area for future investigations.

Concluding Remarks

Despite being described as "docile" in intraspecific groups (Goodpaster and Hoffmeister 1954, Linzey and Packard 1977), few empirical studies have determined the precise nature of interactions between *O. nuttalli* and other small mammal species. Many studies offer observations of golden mice living in close proximity to other small mammal species (Dueser and Shugart 1978, Kitchings and Levy 1981, Linzey 1968, McCarley 1958, Packer and Layne 1991, Pearson 1954). A few investigators concluded that niche partitioning allowed the coexistence of golden mice with similar sympatric species (Feldhamer and Maycroft 1992, Frank and Layne 1992, Knuth and Barrett 1984). It has been speculated, for example, that golden mice select different nest sites and materials (Frank and Layne 1992, Ivey 1949) than other species to provide niche separation. Although these descriptive studies are useful in *identifying* potential interspecific interactions between golden mice and other small mammal species, they offer only limited *evidence* of such relationships.

Removal of potential competitors of golden mice demonstrated that golden mice altered their microhabitat use toward more favorable habitat (Seagle 1985; although see Corgiat 1996) and changed their use of three-dimensional space (Christopher and Barrett 2006). Results such as these support the conclusion of some descriptive studies that interspecific competition exists between golden mice and other small mammal species. It remains unclear, however, whether exploitation or interference competition is the mechanism.

Interactions of golden mice with predators have received little attention. Studies have suggested that use of arboreal nests in dense vegetation (Klein and Layne 1978, Morzillo et al. 2003, Wagner et al. 2000) and well-hidden underground nests (Easterla 1968, Frank and Layne 1992, Packard and Garner 1964) by golden mice are potential adaptations for predator avoidance. Frank and Layne (1992) described weasel predation on *P. gossypinus* but none on golden mice and they attributed this to use of underground nests by golden mice. Future empirical studies are necessary to decipher the role of predation in the ecology of golden mice.

Although some studies have provided evidence of differential rates of parasitism on golden mice compared to other small mammal species (Clark and Durden 2002, Durden of this volume, Jennison et al. 2006), explanations for such differences are speculative. Additional information is needed to better elucidate why golden mice suffer fewer parasites than other species. Such studies would be a welcome contribution to knowledge of the ecology of the golden mouse.

The extensive use of aboveground habitat by golden mice (Christopher and Barrett 2006) suggests that future researchers must view patterns of activity of this species based on three-dimensional use of habitat space to obtain a more realistic picture of interspecific interactions with other small mammal species functioning in the same ecosystem. Finally, it should be noted that the intriguing arboreality of the golden mouse provides mammalogists with a unique opportunity to examine how three-dimensional space use in a species can change in response to habitat alterations, including human-caused forest fragmentation and habitat destruction.

Literature Cited

Andrews, R.D. 1963. The golden mouse in southern Illinois. Natural History Miscellanea 179:1–3.

Barbour, R.W. 1942. Nests and habitat of the golden mouse in eastern Kentucky. Journal of Mammalogy 23:90–91.

Blus, L.J. 1966. Some aspects of golden mouse ecology in southern Illinois. Transactions of the Illinois State Academy of Science 59:334–341.

Burns, C.E., B.J. Goodwin, and R.S. Ostfeld. 2005. A prescription for longer life? Bot fly parasitism of the white-footed mouse. Ecology 86:753–761.

Calhoun, J.B. 1941. Distribution and food habits of mammals in the vicinity of the Reelfoot Lake Biological Station. Journal of the Tennessee Academy of Science 6:207–225.

Carpenter, S.R., and J.F. Kitchell. 1993. The trophic cascade in lakes. Cambridge University Press, Cambridge, United Kingdom.

Christopher, C.C., and G.W. Barrett. 2006. Coexistence of white-footed mice (*Peromyscus leucopus*) and golden mice (*Ochrotomys nuttalli*) in a southeastern forest. Journal of Mammalogy 87:102–107.

Christopher, C.C., and G.W. Barrett. 2007. Double captures of *Peromyscus leucopus* and *Ochrotomys nuttalli*. Southeastern Naturalist 6. In press.

Clark, K.L., and L.A. Durden. 2002. Parasitic arthropods of small mammals in Mississippi. Journal of Mammalogy 83:1039–1048.

Corgiat, D.A. 1996. Golden mouse microhabitat preference: Is there interspecific competition with white-footed mice in southern Illinois? MS thesis, Southern Illinois University, Carbondale, Illinois.

Dayan, T., D. Simberloff, E. Tchernov, and Y. Yom-Tov. 1989. Inter- and intraspecific character displacement in mustelids. Ecology 70:1526–1539.

deRageot, R.H. 1964. The golden mouse. Virginia Wildlife 25:10–11.

Dooley, J.L., Jr, and R.D. Dueser. 1996. Experimental tests of nest site competition in two *Peromyscus* species. Oecologia 105:81–86.

Dueser, R.D., and J.G. Hallett. 1980. Competition and habitat selection in a forest-floor small mammal fauna. Oikos 35:293–297.

Dueser, R.D., and H.H. Shugart, Jr. 1978. Microhabitats in a forest-floor small mammal fauna. Ecology 59:89–98.

Dueser, R.D., and H.H. Shugart, Jr. 1979. Niche pattern in a forest-floor small-mammal fauna. Ecology 60:108–118.

Dunaway, P.B., J.A. Payne, L.L. Lewis, and J.D. Story. 1967. Incidence and effects of *Cuterebra* in *Peromyscus*. Journal of Mammalogy 48:38–51.

Eads, J.H., and J.S. Brown. 1953. Studies on the golden mouse, *Peromyscus nuttalli aureolus*, in Alabama. Journal of the Alabama Academy of Sciences 25:25–26.

Easterla, D.A. 1968. Terrestrial home site of golden mouse. American Midland Naturalist 79:246–247.

Faust, B.F., M.H. Smith, and W.B. Wray. 1971. Distances moved by small mammals as an apparent function of grid size. Acta Theriologica 16:161–177.

Feldhamer, G.A., and K.A. Maycroft. 1992. Unequal capture responses of sympatric golden mice and white-footed mice. American Midland Naturalist 128:407–410.

Feldhamer, G.A., and C.R. Paine. 1987. Distribution and relative abundance of the golden mouse (*Ochrotomys nuttalli*) in Illinois. Transactions of the Illinois Academy of Science 80:213–220.

Frank, P.H., and J.N. Layne. 1992. Nests and daytime refugia of cotton mice (*Peromyscus gossypinus*) and golden mice (*Ochrotomys nuttalli*) in south-central Florida. American Midland Naturalist 127:21–30.

Fuller, A.C., and A.R. Blaustein. 1996. Effects of the parasite *Eimeria arizonensis* on survival of deer mice (*Peromyscus maniculatus*). Ecology 77:2196–2202.

Gause, G.F. 1934. The struggle for existence. Williams and Wilkins Co., Baltimore, Maryland.

Glynn, P.W. 1983. Crustacean symbionts and the defense of corals: Coevolution of the reef? Pages 111–178 *in* M.H. Nitecki, editor. Coevolution. University of Chicago Press, Chicago, Illinois.

Goodpaster, W.W., and D.F. Hoffmeister. 1954. Life history of the golden mouse, *Peromyscus nuttalli*, in Kentucky. Journal of Mammalogy 35:16–27.

Gotelli, N.J., and D.J. McCabe. 2002. Species co-occurrence: A meta-analysis of J.M. Diamond's assembly rules model. Ecology 83:2091–2096.

Hanski, I., H. Henttonen, E. Korrpimaki, I. Oksanen, and P. Turchin. 2001. Specialist predators, generalist predators and the microtine rodent cycle. Ecology 82:1505–1520.

Hurlbert, S.H. 1984. Pseudoreplication and the design of ecological field experiments. Ecological Monographs 54:187–211.

Ivey, R.D. 1949. Life history notes on three mice from the Florida east coast. Journal of Mammalogy 30:157–162.

Jennison, C.A., L.R. Rodas, and G.W. Barrett. 2006. *Cuterebra fontinella* parasitism on *Peromyscus leucopus* and *Ochrotomys nuttalli*. Southeastern Naturalist 5:157–164.

Jewell, M.A., M.K. Anderson, and G.W. Barrett. 1991. Bioenergetics of the golden mouse on experimental sumac seed diets. American Midland Naturalist 125:360–364.

Kennedy, M.L., J.W. Hardin, and M.J. Harvey. 1973. Circadian rhythm in the golden mouse, *Ochrotomys nuttalli*. Journal of the Tennessee Academy of Science 48:77–79.

Kitchings, J.T., and D.J. Levy. 1981. Habitat patterns in a small mammal community. Journal of Mammalogy 62:814–820.

Klein, H.G., and J.N. Layne. 1978. Nesting behavior in four species of mice. Journal of Mammalogy 59:103–108.

Knuth, B.A., and G.W. Barrett. 1984. A comparative study of resource partitioning between *Ochrotomys nuttalli* and *Peromyscus leucopus*. Journal of Mammalogy 65:576–583.

Kronfeld-Schor, N., and T. Dayan. 2003. Partitioning of time as an ecological resource. Annual Review of Ecology and Systematics 34:153–181.

Layne, J.N. 1958. Notes on mammals of southern Illinois. American Midland Naturalist 60:219–254.

Linzey, D.W. 1968. An ecological study of the golden mouse, *Ochrotomys nuttalli*, in the Great Smoky Mountains National Park. American Midland Naturalist 79:320–345.

Linzey, D.W., and R.L. Packard. 1977. *Ochrotomys nuttalli*. Mammalian Species 75:1–6.

Maehr, D.S., and J.R. Brady. 1986. Food habits of bobcats in Florida. Journal of Mammalogy 67:133–138.

Maser, C., J.M. Trappe, and R.A. Nussbaum. 1978. Fungal–small mammal interrelationships with emphasis on Oregon coniferous forests. Ecology 59:799–809.

McCarley, H. 1958. Ecology, behavior and population dynamics of *Peromyscus nuttalli* in eastern Texas. Texas Journal of Science 10:147–171.

McKenzie, N.L., and J.K. Rolfe. 1986. Structure of bat guilds in the Kimberley mangroves, Australia. Journal of Animal Ecology 55:401–420.

Morzillo, A.T., G.A. Feldhamer, and M.C. Nicholson. 2003. Home range and nest use of the golden mouse (*Ochrotomys nuttalli*) in southern Illinois. Journal of Mammalogy 84:553–560.

Odum, E.P., and G.W. Barrett. 2005. Fundamentals of ecology, 5th ed. Thompson Brooks/Cole, Belmont, California.

O'Malley, M., J. Blesh, M. Williams, and G.W. Barrett. 2003. Food preferences and bioenergetics of the white-footed mouse (*Peromyscus leucopus*) and the golden mouse (*Ochrotomys nuttalli*). Georgia Journal of Science 61:233–237.

Ostfeld, R.S., and K.M. LoGiudice. 2003. Community disassembly, biodiversity loss, and the erosion of an ecosystem service. Ecology 84:1421–1427.

Packard, R.L., and H. Garner. 1964. Arboreal nests of the golden mouse in eastern Texas. Journal of Mammalogy 45:369–374.

Packer, W.C., and J.N. Layne. 1991. Foraging site preferences and relative arboreality of small rodents in Florida. American Midland Naturalist 125:187–194.

Pearson, P.G. 1954. Mammals of Gulf Hammock, Levy County, Florida. American Midland Naturalist 51:468–480.

Peles, J.D., C.K. Williams, and G.W. Barrett. 1995. Bioenergetics of golden mice: The importance of food quality. American Midland Naturalist 133:373–376.

Power, M.E. 1992. Top-down and bottom-up forces in food webs: Do plants have primacy? Ecology 73:733–746.

Power, M.E., D. Tilman, J.A. Estes, B.A. Menge, W.J. Bond, L.S. Mills, G. Daily, J.C. Castilla, J. Lubchenco, and R.T. Paine. 1996. Challenges in the quest for keystones. BioScience 46:609–620.

Pruett, A.L., C.C. Christopher, and G.W. Barrett. 2002. Effects of a forested riparian peninsula on mean home range size of the golden mouse (*Ochrotomys nuttalli*) and the white-footed mouse (*Peromyscus leucopus*). Georgia Academy of Science 60:201–208.

Ray, J.C., and M.E. Sunquist. 2001. Trophic relations in a community of African rainforest carnivores. Oecologia 127:395–408.

Schmitz, O., and T. Nudds. 1994. Parasite-mediated competition in deer and moose: How strong is the effect of meningeal worm on moose? Ecological Applications 41:91–103.

Schoener, T.W. 1974. Resource partitioning in ecological communities. Science 185:27–39.

Shugart, H.H., and B.C. Patten. 1972. Niche quantification and the concept of niche pattern. Pages 284–327 *in* B.C. Patten, editor. Systems analysis and simulation in ecology, Vol. 2. Academic Press, New York, New York.

Seagle, S.W. 1985. Competition and coexistence of small mammals in an east Tennessee pine plantation. American Midland Naturalist 114:272–282.

Stachowicz, J. 2001. Mutualism, facilitation, and the structure of ecological communities. BioScience 51:235–246.

Wagner, D.M., G.A. Feldhamer, and J.A. Newman. 2000. Microhabitat selection by golden mice (*Ochrotomys nuttalli*) at arboreal nest sites. American Midland Naturalist 144:220–225.

Wolfe, J.L., and A.V. Linzey. 1977. *Peromyscus gossypinus*. Mammalian Species 70:1–5.

5
Ecosystem Ecology of the Golden Mouse

STEVEN W. SEAGLE

The golden mouse, Ochrotomys nuttalli, described by Harlan in 1832, has remained one of the least studied members of the fauna of the southeastern United States– (Linzey 1968:320)

Role of the Golden Mouse in Ecosystem Dynamics

Various mammalian examples of "dominant species" such as the white-tailed deer (*Odocoileus virginianus*; Seagle 2003), "keystone species" such as the blue wildebeest (*Connochaetes taurinus*; McNaughton et al. 1988), and "ecological engineers" such as the beaver (*Castor canadensis*; Naiman et al. 1994) have been described, clearly representing species that strongly influence various aspects of ecosystem structure and function. Small rodents have also been noted to strongly affect pattern and change in ecosystems. For example, Stephen's kangaroo rats (*Dipodomys stephensi*) influence the composition and spatial pattern of plant communities through their burrowing and foraging activities (Brock and Kelt 2004). Similarly, black-tailed prairie dogs (*Cynomys ludovicianus*) create hot spots of vertebrate diversity (Lomolino and Smith 2003) and unique vegetation composition and structure that facilitate nutrient and energy flux (Fahnestock and Detling 2002). Rodents have long been proposed as a significant vector for dissemination of fungal spores in forests of the Pacific Northwest (Maser et al. 1978). More recent studies continue to highlight the extent of mycophagy in small rodents (Colgan and Claridge 2002, Orrock and Pagels 2002) and the potential for consumption and passage through small mammals to enhance the germination and the inoculation potential of spores (Caldwell et al. 2005). Based on various studies, Inouye et al. (1987) suggested a positive feedback among arvicoline rodent consumption of plant material, subsequent deposition of nutrients in fecal matter or urine, and regrowth of food plants. Pastor et al. (1996) found that the total amounts of nitrogen and phosphorus mineralized by meadow voles (*Microtus pennsylvanicus*) and red-backed voles (*Clethrionomys gapperi*) were relatively small components of the nutrient budgets for a 13-year-old forest stand. They did note, however, that localized fertilization and spore dispersal to microsites of seedling regeneration might significantly impact forest dynamics.

In contrast to localized impacts, Clark et al. (2005) estimated that rodents in an Oklahoma old field deposit fecal and urinary nitrogen in amounts comparable to large herbivores and other system nutrient fluxes and thus represent an integral part of the terrestrial old-field nitrogen cycle. This list of small mammal interactions with ecosystem structure and function is highly abbreviated (Hayward and Phillipson 1979), but it does highlight the variety and extent of small mammal impacts on ecosystems and prompt the question: "What is the role of the golden mouse (*Ochrotomys nuttalli*) in forest ecosystems?"

"Passive" Versus "Active" Species

To classify the roles of animal species in forest dynamics, MacMahon (1981) posed two basic questions: (1) How do specific animal species affect ecosystem components and subsequently have an impact on succession? (2) How can succession of various ecosystem components affect animals? These questions describe a simplistic, but heuristically useful, dichotomy of the role that animals might play in ecosystems. Are they "active participants" in ecosystem function and change, or are they "passive responders" to the dynamics of food, habitat/microhabitat, and other resources represented by ecosystems? To my knowledge, no scientific studies have been carried out to specifically examine the impacts that golden mice might have in forest ecosystem structure or function. Thus, the tacit assumption has been that golden mice are simply "passive responders" to forest ecosystem changes, rather than "ecosystem engineers" (Wright and Jones 2006; but see Chapter 9 of this volume). Such an assumption is logically derived for a species that normally occurs at low density (Kitchings and Levy 1981, Linzey 1968, Chapters 3 and 7 of this volume), maintains a scattered spatial distribution that is highly dependent on microhabitat structure (Linzey 1968), consistently is characterized as having a narrow niche breadth (Dueser and Shugart 1979, Seagle 1985a), often assumes a subservient role with sympatric species (Seagle 1985b; but see also Christopher and Barrett 2006), occupies forest habitats where most biomass is held in recalcitrant aboveground (tree biomass) and belowground (tree roots and the "slow" fraction of soils) pools, and, at most, nominally feeds on the relatively labile nutrients and biomass made available in annual leaf production. Consequently, the primary potential for golden mice to impact forest ecosystems seems to be through feeding preferences on fruits, seeds, and insects.

Potential Trophic Roles of Golden Mice

Early accounts of golden mouse food habits established that the species is omnivorous (Calhoun 1941, Goodpaster and Hoffmeister 1954, Linzey 1968), feeding primarily on seeds, fruits, and invertebrates. This diet breadth and composition opens opportunity for the species to influence forest ecosystems through seed predation, seed dispersal, and predation of key invertebrate taxa.

A comparison of diets in different habitats/microhabitats suggests that seeds and fruits from a variety of trees, shrubs, and vines are eaten, with relative use varying according to local availability. For example, Linzey (1968) found that greenbrier (*Smilax glauca*) seeds were very common in stomach contents at sites having plentiful greenbrier in the understory. However, in sites having less greenbrier or species of *Rubus*, the amount of insect material in the stomach increased. This opportunistic pattern of foraging does not suggest a close trophic association with specific plant species and might indicate a low likelihood of golden mice having a significant impact on the reproduction and population dynamics of any single plant species. Of course, even opportunistic feeding can impact population dynamics of plants that are rare. However, the tree species (representing the genera *Prunus*, *Cornus*, *Quercus*, *Robinia*, and *Liriodendron*) and vine or shrub species (representing the genera *Smilax*, *Rhus*, *Rubus*, *Lonicera*, and *Vitis*) that have been specifically noted to contribute to the golden mouse diet (Goodpaster and Hoffmeister 1954, Linzey 1968) are generally not rare.

Despite various general descriptions of the golden mouse diet, concurrent quantitative examinations of relative consumption and food availability do not seem to have been conducted. This places a significant limitation on our ability to interpret potential interactions or develop hypotheses regarding the role of golden mice in existing or changing plant communities. One exceptional example might be that seeds of the genus *Cornus* are often noted in studies of golden mouse food habits and observations (G.W. Barrett, personal communication), suggesting that seeds of flowering dogwood (*Cornus florida*) are a highly favored food source. Because populations of flowering dogwood are currently being devastated by dogwood anthracnose (caused by *Discula destructiva*) over much of its range, the potential for preferential feeding by golden mice and other rodents might put further negative pressure on dwindling dogwood populations (Figure 5.1). These combined negative effects on flowering dogwood could effectively offset the species as a "calcium pump" (Day and McGintry 1975, Thomas 1969) in southeastern forest ecosystems, thereby reducing calcium availability to many biotic components of the forest. Obviously, detrimental effects on golden mouse populations could also ensue from the decline of flowering dogwood. When articulated, such pieces of information provide tantalizing opportunities for further study, including the rich diversity of such interactions that might be uncovered by a more complete description of golden mouse feeding habitats relative to food availability.

Seed Dispersal

Neither seed nor spore dispersal has been intensively or quantitatively studied for the golden mouse. Clearly, the abundance of seeds in golden mouse nests (Goodpaster and Hoffmeister 1954, Linzey 1968) and their consumption of acorns (*Quercus* spp.; Christopher and Barrett 2006, Linzey 1968, O'Malley et al. 2003)

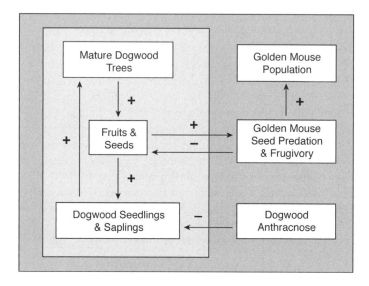

FIGURE 5.1. The flowering dogwood (*C. florida*) reproductive cycle is being disrupted by disease (dogwood anthracnose; Jenkins and White 2002) but might also be negatively impacted by golden mouse frugivory and seed predation. Anthracnose alone is causing dogwood population declines and golden mice might exacerbate this decline. As a primary food resource, dogwood decline will negatively impact golden mouse populations.

allow the possibility for this species to disperse plant propagules to either favorable or unfavorable sites of germination. Interestingly, mycophagy is not noted in existing golden mouse food studies even though the omnivorous nature of the species and fungi consumption by similar, sympatric mouse species (Orrock et al. 2003) offer the possibility that golden mice play a significant role in spore dispersal. In addition, it has been suggested that the distribution and availability of fungi can affect habitat selection by mice (Brannon 2005) and that interspecific competition for fungi might occur (Orrock et al. 2003). Mycophagy might be limited if golden mice truly feed very little on the forest floor (Goodpaster and Hoffmeister 1954), but this assertion seems questionable given the abundance of studies that have reported golden mice trapped on the forest floor.

Christopher and Barrett (2006) suggested that mast cropping by water oak (*Quercus nigra*) can dampen competition between golden mice and white-footed mice (*Peromyscus leucopus*). This suggestion is consistent with the theory that mast cropping evolved to satiate seed predators and ensure periodic survival of plant propagules (Kelly and Sork 2002). The corollary to this theory is that the collective consumption of acorns by golden mice and numerous other species of seed predators in nonmast years has placed selective pressure on oak reproductive strategies. Parceling out this pressure among species to determine the importance of golden mice is not feasible without community-level seed production and predation data (e.g., Schnurr et al. 2002). Such studies might yield interesting insights, particularly for habitats where golden mice comprise a significant component of the seed predator community and where plant species favored by golden mice (e.g., *C. florida* and *Q. nigra*) are common.

Invertebrate Foods

Lack of details regarding the species and proportion of standing crop of invertebrates consumed by the golden mouse makes it speculative to assess the effect of this species in forest trophic cascades. Calhoun (1941) identified Araneida, Coleoptera, and Lepidoptera as components of the golden mouse diet in Tennessee. Lepidopteran larvae are important consumers of green leaf biomass in forest habitats (Gosz et al. 1978, Rinker et al. 2001) and the Araneida play a pivotal role as predators in the detrital food web of the forest floor (Chen and Wise 1999). Thus, any significant predation impact on key species or populations of these taxa by golden mice could exert a strong top-down effect on both carbon and nutrient cycles.

Microhabitat and Diet

Some of the more intriguing suggestions regarding the trophic role of golden mice involve the interaction of habitat/microhabitat use and feeding bioenergetics (Christopher and Barrett 2006, O'Malley et al. 2003, Peles et al. 1995). These studies show that after removal of sympatric white-footed mice, golden mice did not change in abundance. They did expand three-dimensional use of space, particularly making greater use of the forest floor where Chinese privet (*Ligustrum sinense*) provided both cover and food (Christopher and Barrett 2006). In an experimental feeding study, O'Malley et al. (2003) found that golden mice strongly preferred water oak acorns over privet berries and suggested that use of privet thickets was primarily for nesting sites and for predator avoidance. Based on body weight maintenance of captive golden mice, Peles et al. (1995) suggested a strong positive relationship among microhabitat use for nesting, predator avoidance, and feeding for golden mice using thickets of Japanese honeysuckle (*Lonicera japonica*). In comparison, a diet of berries from the Eastern red cedar (*Juniperus virginiana*) resulted in loss of body mass by captive golden mice, even though red cedar is commonly used for nest sites (Andrews 1963, Morzillo et al. 2003, Stueck et al. 1977). These results help explain the lower abundance of golden mice in cedar glades compared to unmanaged pine plantations having abundant honeysuckle (Seagle 1985a, 1985b). They also suggest a trophic basis for habitat selection by the species and define a strong trophic linkage between golden mice and honeysuckle in forest ecosystems for future study.

Forest Habitat Disturbance, Succession, and Management

Golden mice inhabit a variety of forest types. However, as noted by Rose (Chapter 3 of this volume), long-term demographic studies of golden mice are uncommon and densities are often low. Consequently, the population and metapopulation responses of golden mice to changing forest ecosystems and patchiness must be inferred from the types of habitat occupied and changes in microhabitat likely to occur during ecosystem disturbance and secondary succession. Fortunately, because of its microhabitat requirement, the golden mouse represents a model species for the study of habitat selection at spatial scales ranging from microhabitat to the landscape (see Chapter 1 of this volume). Seagle (1983) captured the hierarchical nature of habitat use for the species by sampling the small mammal fauna found in a variety of habitats on the Oak Ridge National Environmental Research Park in Tennessee. By trapping in forest and old-field habitats of various types and ages and measuring the same suite of habitat/microhabitat characteristics at each capture site, he performed a principle components analysis that reflected selectivity of key habitats and microhabitats (Figure 5.2).

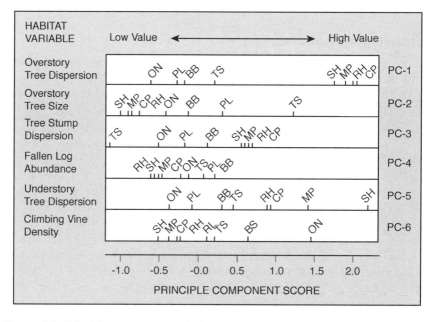

FIGURE 5.2. Principle components analysis displays the relationships of eight small mammal species to habitat and microhabitat structural characteristics in a hierarchical manner from gross habitat to specific microhabitat variables. Placement of each species along each principle component represents the mean score for the species. Abbreviations: ON, golden mouse; PL, white-footed mouse; BB, short-tailed shrew; TS, Eastern chipmunk; SH, cotton rat; MP, pine vole; RH, eastern harvest mouse; CP, least shrew.

Principle component 1 (Figure 5.2) clearly separated the old-field species [cotton rat (*Sigmodon hispidus*), pine vole (*Microtus pinetorum*), eastern harvest mouse (*Reithrodontomys humulis*), and least shrew (*Cryptotis parva*)] from the forest-dwelling species [golden mouse, white-footed mouse, and Eastern chipmunk (*Tamias striatus*)] and the ubiquitous northern short-tailed shrew (*Blarina brevicauda*). This principle component demonstrates the patchy use of the landscape by golden mice through the selection of forest habitats. Of the forest species examined, the golden mouse used stands with smaller overstory trees (principle component 2), indicating preference for younger forest stands. Use of microhabitats characterized by lower dispersion of tree stumps (principle component 3) and higher density of understory trees (principle component 5) suggests that golden mice use either disturbed sites within forests or young forest stands that have not undergone a high degree of self-thinning. Finally, the golden mouse stands out on principle component 6 as the species most dependent on the presence of climbing vines such as Japanese honeysuckle and poison ivy (*Toxicodendron radicans*) for suitable microhabitat (Figure 5.2). These habitat/microhabitat characteristics of golden mice were manifest in the capture of most individuals in relatively young or unmanaged pine plantations. From a geographical perspective, Perry and Thill (2005) also reported the association of golden mice with moderate to dense woody understory in pine regeneration sites found in the Quachita Mountains in Arkansas and Oklahoma. Primary use of early or mid-successional forests or managed forests for appropriate habitat describes a structural niche that is dependent on disturbance, intensity of forest management practices, and the presence of at least one prominent exotic species, namely Japanese honeysuckle.

Upland Forests

Both human and natural disturbances are integral in creating the habitat mosaic used by golden mice in mature upland forests of the southeastern United States (Dueser and Shugart 1979, Kitchings and Levy 1981, Seagle 1985a). At small spatial scales, these forests are characterized by the relatively frequent creation of canopy gaps where light conditions stimulate rapid development of dense understory trees intermixed with woody shrubs and vines. Such events create three-dimensional habitat/microhabitat for golden mice but might also represent relatively small and potentially isolated patches of suitable microhabitat. Colonization and persistent occupation of such patches would likely depend on distance to source populations as well as the spatial and temporal pattern of patch creation (see Chapter 6 of this volume).

Drier upland hardwood or mixed conifer-hardwood sites are also subject to fire in addition to gap-phase forest succession (Oliver and Larson 1990). Low-intensity ground fires in these forests might diminish the shrubby understory microhabitat for golden mice. This certainly alters the temporal dynamics of appropriate microhabitat and perhaps decreases the likelihood of occupation completely.

Although widespread and long-term studies of golden mouse populations are lacking to confirm it, decades of fire suppression by humans (Sharitz et al. 1992) could have stabilized golden mouse habitat in time and even expanded its extent. Growing emphasis on implementing "natural" fire disturbance regimes might have the opposite effect by removing vines and debris from the understory. Stand-leveling disturbances, such as windstorms, crown fires, or tornados (Loeb 1999), frequently create large patches of appropriate habitat for golden mice where shading by extensive vine and shrub growth also helps to delay succession. Nonetheless, except for stand edges, eventual canopy closure and stand self-thinning would diminish golden mouse habitat.

Mesic Forests

Mixed mesic upland and riparian, or bottomland, forests are also occupied by golden mice (Christopher and Barrett 2006, Jennison et al. 2006, McCarley 1958, Schmid-Holmes and Drickamer 2001, Smith et al. 1974). With wetter conditions, fire is less likely to occur as a disturbance, thus extensive growth of canebrakes, greenbrier, and honeysuckle provides understory and shrub-level characteristics conducive to golden mouse occupation. Gap-phase succession in these forests is less likely to create unique patches of microhabitat for golden mice but certainly can enhance the existing habitat. Although speculative, this forest type might well provide the most temporally stable and spatially continuous forest habitat used by golden mice.

Pine Plantations

Pine forests, particularly managed pine plantations, might be even more predictable in terms of habitat quality for golden mice. Atkeson and Johnson (1979) studied mammal community succession on 32 pine plantations ranging from 1 to 15 years of age on the Georgia Piedmont. These plantations were established using intensive site preparation techniques, including clear-cutting, felling of all remnant trees, raking of slash and excavated roots into windrows and burning, and disking of the cleared ground before planting. The resulting secondary succession of herbaceous plants and pines produced dense patches of blackberry (*Rubus* spp.) in 5 years. After 7 years, crown closure was beginning and honeysuckle was noted in most patches. Golden mice were evident in low densities when plantations reached 5 years and remained present in all older plantations. This study clearly defined the negative impact of intensive site preparation for pine plantations on golden mice, as well as the progression of habitat recovery during secondary succession.

Constantine et al. (2004) examined the effect of retaining pine corridors when pine plantations are harvested. They found that golden mice were present (although not abundant) in original pine plantations and in the 100-m-wide corridors left after harvesting, but they were not present in the harvested area. These plantations were 20–23 years old, with scattered hardwoods in the midstory and an understory of wild grape (*Vitis* spp.), greenbrier, poison ivy, and Virginia creeper (*Parthenocissus quinquefolia*). This plant community composition and structure resulted from intensive site preparation (sheering, root raking, and bedding for planting), but no further management had been applied. Mengak and Guynn (2003) examined microhabitat use by small mammals in loblolly pine (*Pinus taeda*) plantations that were naturally regenerated (seed tree regeneration) or prepared for planting (shearing, chopping, and burning). Although apparently found in both types of plantation when appropriate microhabitat structure was available, they noted that those trapping stations that yielded golden mice were mainly in naturally regenerated stands.

Perry and Thill (2005) examined four shortleaf pine (*Pinus echinata*) regeneration treatments in Arkansas and Oklahoma. Golden mice capture rates were highest in shelterwood cuts sampled up to 5 years after harvest (0.77 captures/100 trap nights) compared to unharvested stands (0.28 captures/100 trap nights). Unharvested stands were mature (≥ 60 years old). Second-growth pine-hardwoods had little herbaceous and woody understory vegetation, whereas shelterwood cuts had dense woody understory vegetation. In examining the small mammal community of loblolly pine plantations ranging in age from 1 to 24 years, Dolan and Rose (2007) found that the golden mouse was consistently present in stands after crown closure at 8 years of age, although the capture rate for the species declined slowly following canopy closure.

Pine plantations in general are viable, dynamic habitats for golden mice. Even though prescribed fire can decrease habitat quality for small mammals in some ecosystems (e.g., Converse et al. 2006), it is interesting that few studies of controlled burns in plantations have noted effects on golden mice. Furtak-Maycroft (1991) examined the small mammal fauna on 21 shortleaf and loblolly pine stands throughout Shawnee National Forest in southern Illinois. All stands had been selectively cut and burned from 1 to 48 years prior to livetrapping for mammals. Golden mice occurred on 14 of the stands, spanning the entire range of time since burning. Although prescribed fire apparently did not remove all appropriate microhabitat in that study, it seems very likely that burns intensive enough to mitigate hardwood encroachment in the understory would be detrimental to the golden mouse microhabitat. Consequently, studies of intensity and frequency of prescribed fire impacts on the golden mouse microhabitat would be useful to identify levels of this disturbance that are beneficial to forest management and also tolerable for golden mice. As greater attention is given to forest plantations as ecological entities in the landscape (Sharitz et al. 1992), such information will be integral to appropriate management. Nonetheless, plantations remain an ephemeral habitat because they are usually harvested within 30–40 years of planting. Consequently, even though occupied early after establishment, use of pine plantation habitat by the golden mouse is a question of patch dynamics and metapopulation dynamics at the landscape scale (see Chapter 6 of this volume).

Golden Mice and Global Change

Whether golden mice are active drivers of any process in forest ecosystems or a passive responder to ecosystem dynamics, multiple aspects of global change have interesting implications for the future of this species. Of the myriad of changes in southeastern forest ecosystems that might result from human activities, the effects of (1) invasive species, and (2) increasing atmospheric CO_2 might affect the golden mouse by impacts on habitat, microhabitat, food resources, trophic dynamics, and species interactions.

Future Range Expansion?

The current range of the golden mouse covers much of the southeastern United States (Feldhamer and Linzey Chapter 2 of this volume, Linzey and Packard 1977, Packard 1969) and includes areas represented by multiple dominant forest types (Iverson and Prasad 2001): oak-hickory, oak-pine, loblolly-slash pine, longleaf-slash pine, and oak-gum-cypress. Projections of these forest-type distributions following doubling of atmospheric CO_2 vary depending on which general circulation model is used to predict climate changes. Nonetheless, results from five different models (Iverson and Prasad 2001) all clearly indicate an expansion northward of the area covered by these forest types collectively. Although proportions of these forest types across the lower Southeast would change, no major loss of forest types that could be inhabited by golden mice is apparent. The net effect would be a significant expansion of the forest types currently within the range of the golden mouse, with a particularly strong expansion of oak-hickory and oak-pine (Figure 5.3; Iverson and Prasad 2001). Assuming that golden mice would react to physical environmental changes in manners similar to tree species, they could experience a large northward range expansion under common scenarios of climate change. Consequently, this species could be an interesting candidate for quantitative niche modeling using multivariate clustering of current and projected environmental variables (Hargrove and Hoffman 2005). Given the patchy distribution of golden mice based on microhabitat structure, various species of climbing vines might also need to respond in a similar manner for this range extension to be realized.

Microhabitat and Elevated CO_2

The question of how the primary microhabitat requirements of the golden mouse might change under elevated CO_2 conditions has been at least partially answered. There is growing worldwide evidence that elevated atmospheric CO_2 might stimulate primary production of woody climbing vines (Phillips et al. 2002) because of their potential to allocate a greater proportion of CO_2-stimulated photosynthate to the production of photosynthetic tissue rather than support tissue (Mohan et al. 2006). Free-air CO_2 enrichment (FACE) studies at Oak Ridge National

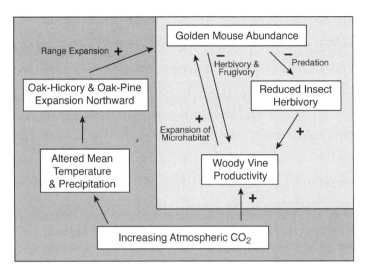

FIGURE 5.3. With predicted climate change from increasing atmospheric greenhouse gases, two forest types used by the golden mouse, oak-hickory and oak-pine, might expand northward, resulting in a possible range expansion for the species. Golden mouse trophic interactions (light gray inset) might also be affected by increasing atmospheric CO_2 through higher productivity of woody vines, which constitute a major microhabitat requirement. The complexity of these trophic interactions, including additional direct and indirect effects of climate change, will be a major challenge for understanding the trophic ecology of the golden mouse.

Laboratory specifically found that Japanese honeysuckle experienced 2.5–3.3 times normal net annual primary production under elevated CO_2 during both wet and dry years (Belote et al. 2003). Mohan et al. (2006) also found that under elevated CO_2, poison ivy increased photosynthesis, water use efficiency, growth, and population biomass relative to ambient conditions. Both of these experimental results occurred under field conditions. Thus, even before physical driving variables shift enough to alter the ranges of long-lived tree species, key vine species providing microhabitat for the golden mouse might display increased productivity and expansion of microhabitat for golden mice (Figure 5.3).

Golden Mice and Invasive Plant Species

Weltzin et al. (2003) raised the possibility that elevated atmospheric CO_2 levels might facilitate non-native plant invasion of existing plant communities. Japanese honeysuckle, being a non-native species, is one example of such potential. As a favored food source (Peles et al. 1995) of the golden mouse and an established cornerstone of its microhabitat, the spread of Japanese honeysuckle has clearly benefited the golden mouse. It is unclear whether all invasive vine and shrub species will benefit from changes in atmospheric CO_2. Regardless, multiple invasive species are currently changing the microhabitat structure of forest habitats and golden mouse habitat/microhabitat. For example, Christopher and Barrett (2006) found that Chinese privet altered microhabitat structure and increased potential food resources for the golden mouse in Georgia and possibly altered interspecific interactions with the white-footed mouse. Other shrubby exotic species, such as the bush honeysuckles (*Lonicera* spp.) and Japanese barberry (*Berberis thunbergii*) might provide similar but as yet unrecognized ecosystem services to the golden mouse.

Conclusions

Although it seems astonishing for such a recognizable and charismatic species as the golden mouse, the opening statement of the Linzey (1968) landmark paper on the species still rings true regarding the role that the golden mouse plays in forest ecosystems: "The golden mouse, *Ochrotomys nuttalli*, described by Harlan in 1832, has remained one of the least studied members of the fauna of the southeastern United States." This sentiment is not repeated here to minimize the numerous studies that have examined golden mouse taxonomy, reproduction, bioenergetics, habitat/microhabitat ecology, behavior, and coevolutionary relationships. Rather, it is meant to encourage reevaluation of the golden mouse as an entity of study, and creativity in view of the roles that the species might play within forest ecosystems.

The studies reviewed in this chapter uncover no "smoking gun" that identifies the golden mouse as a dominant or keystone species in forest ecosystems. Consequently, the species might be viewed as a "passive responder" to the changes in habitat and food brought about by other factors, such as human management of forest ecosystems. This conclusion might also be premature. For example, without a better understanding of golden mouse feeding habits relative to food availability, it seems rash to summarily classify the species as a minor consumer of berries and insects. The potential interactions among golden mice, flowering dogwood, and dogwood anthracnose represent but one example of possible impacts golden mice might have on trophic interactions.

The close trophic and microhabitat association of the golden mouse with Japanese honeysuckle takes on new dimensions if elevated atmospheric CO_2 simultaneously increases production of honeysuckle (including fruits) and suitable golden mouse habitat space. Could such a synergy broadly increase the densities of golden mouse populations, making the species a new indicator of global change in southeastern forest ecosystems? In addition, if golden mice are a significant predator of invertebrates that feed on honeysuckle, might golden mice provide further positive feedback (Figure 5.3) for the expansion of honeysuckle and its negative impacts on tree regeneration? This latter question will remain unanswerable without greater detail of golden mouse food preferences.

Although the golden mouse is distributed patchily at the landscape scale and often found at low density, these characteristics are in common with most other species on Earth (see Chapter 7 of this volume). Thus, rather than allowing such demographic characteristics to discourage study, the golden mouse should represent opportunity for insight into the ecology of numerous species. Although the effort necessary to perform field studies of small mammals is well known, the golden mouse is readily identifiable, it is relatively easy to capture, its basic reproductive biology is documented, and its habitat and microhabitat are quite recognizable. Collectively, these characteristics make the golden mouse remarkably well suited for studying the ecology of a relatively rare species through dynamic habitat modeling and new techniques such as quantitative niche modeling.

The role that biodiversity plays in ecosystem function will continue to be debated (Chapin et al. 2000), but only further study of the natural history, population ecology, interspecific interactions, and trophic impacts of the less common species in ecosystems will actually elucidate the biological and economic importance of biodiversity. The golden mouse represents a conspicuous and compelling component of biodiversity in southeastern forest ecosystems, and without additional study, it remains premature to make final judgments on its role in forest ecosystems.

Acknowledgments. J.D. Dolan and R.K. Rose provided access to an unpublished manuscript. G.W. Barrett and G.A. Feldhamer provided continual encouragement and support during the writing. Conceptualization and development of this work was partially supported by grant DEB-0454600 from the National Science Foundation to Appalachian State University and S.W. Seagle.

Literature Cited

Andrews, R.D. 1963. The golden mouse in southern Illinois. Chicago Academy of Sciences Natural History Miscellaneous Publication 179:1–3.

Atkeson, T.D., and A.S. Johnson. 1979. Succession of small mammals on pine plantations in the Georgia Piedmont. American Midland Naturalist 101:385–392.

Belote, R.T., J.F. Weltzin, and R.J. Norby. 2003. Response of an understory plant community to elevated $[CO_2]$ depends on differential responses of dominant invasive species and is mediated by soil water availability. New Phytologist 161:827–835.

Brannon, M.P. 2005. Distribution and microhabitat of the woodland jumping mouse, *Napaeozapus insignis*, and the white-footed mouse, *Peromyscus leucopus*, in the Southern Appalachians. Southeastern Naturalist 4:479–486.

Brock, R.E., and D.A. Kelt. 2004. Keystone effects of the endangered Stephen's kangaroo rat (*Dipodomys stephensi*). Biological Conservation 116:131–139.

Caldwell, I.R., K. Vernes, and F. Barlocher. 2005. The northern flying squirrel (*Glaucomys sabrinus*) as a vector for inoculation of red spruce (*Picea rubens*) seedlings with ectomycorrhizal fungi. Sydowia 57:166–178.

Calhoun, J.B. 1941. Distribution and food habits of mammals in the vicinity of the Reelfoot Lake Biological Station. Journal of the Tennessee Academy of Science 6:177–185, 207–225.

Chapin, F.S., E.S. Zavaleta, V.T. Eviner, R.L. Naylor, P.M. Vitousek, H.L. Reynolds, D.U. Hooper, S. Lavorel, O.E. Sala, S.E. Hobbie, M.C. Mack, and S. Diaz. 2000. Consequences of changing biodiversity. Nature 405:234–242.

Chen, B., and D.H. Wise. 1999. Bottom-up limitation of predaceous arthropods in a detritus-based terrestrial food web. Ecology 80:761–772.

Christopher, C.C., and G.W. Barrett. 2006. Coexistence of white-footed mice (*Peromyscus leucopus*) and golden mice (*Ochrotomys nuttalli*) in a southeastern forest. Journal of Mammalogy 87:102–107.

Clark, J.E., E.C. Hellgren, J.L. Parsons, E.E. Jorgensen, D.M. Engle, and D.M. Leslie. 2005. Nitrogen outputs from fecal and urine deposition of small mammals: Implications for nitrogen cycling. Oecologia 144:447–455.

Colgan, W., and A.W. Claridge. 2002. Mycorrhizal effectiveness of *Rhizopogon* spores recovered from fecal pellets of small forest-dwelling mammals. Mycological Research 106:314–320.

Constantine, N.L., T.A. Campbell, W.M. Baughman, T.B. Harrington, B.R. Chapman, and K.V. Miller. 2004. Effects of clearcutting with corridor retention on abundance, richness, and diversity of small mammals in the Coastal Plain of South Carolina, USA. Forest Ecology and Management 202:293–300.

Converse, S.J., W.M. Block, and G.C. White. 2006. Small mammal population and habitat responses to forest thinning and prescribed fire. Forest Ecology and Management 228:263–273.

Day, F.P., Jr., and D.T. McGintry. 1975. Mineral cycling strategies of two deciduous and two evergreen tree species on a southern Appalachian watershed. Pages 736–743 *in* F.C. Howell, J.B. Gentry, and M.H. Smith, editors. Mineral cycling in southeastern ecosystems. United States Department of Commerce National Technical Information Service, Washington, DC.

Dolan, J.D., and R.K. Rose. 2007. Small mammal communities in pine plantations in eastern Virginia. Journal of the Virginia Academy of Science. In press.

Dueser, R.D., and H.H. Shugart, Jr. 1979. Niche pattern in a forest-floor small-mammal fauna. Ecology 60:108–118.

Fahnestock, J.T., and J.K. Detling. 2002. Bison–prairie dog–plant interactions in a North American mixed-grass prairie. Oecologia 132:86–95.

Furtak-Maycroft, K.A. 1991. An empirically-based habitat suitability index model for golden mice, *Ochrotomys nuttalli*, on pine stands in Shawnee National Forest. MS thesis, Southern Illinois University, Carbondale, Illinois.

Goodpaster, W.W., and D.F. Hoffmeister. 1954. Life history of the golden mouse, *Peromyscus nuttalli*, in Kentucky. Journal of Mammalogy 35:16–27.

Gosz, J.R., R.T. Holmes, G.E. Likens, and F.H. Bormann. 1978. The flow of energy in a forest ecosystem. Scientific American 238:92–102.

Hargrove, W.W., and F.M. Hoffman. 2005. Potential of multivariate quantitative methods for delineation and visualization of ecoregions. Environmental Management Supplement S39–S60.

Hayward, G.F., and J. Phillipson. 1979. Community structure and functional role of small mammals in ecosystems. Pages 135–211 *in* D.M. Stoddart, editor. Ecology of small mammals. Chapman & Hall, London, United Kingdom.

Inouye, R.S., N.J. Huntly, D. Tilman, and J.R. Tester. 1987. Pocket gophers (*Geomys bursarius*), vegetation, and soil nitrogen along a successional sere in east-central Minnesota. Oecologia 72:178–184.

Iverson, L.R., and A.M. Prasad. 2001. Potential changes in tree species richness and forest community types following climate change. Ecosystems 4:186–199.

Jenkins, M.A., and P.S. White. 2002. *Cornus florida* L. mortality and understory composition changes in western Great Smoky Mountains National Park. Journal of the Torrey Botanical Society 129:194–206.

Jennison, C.A., L.A. Rodas, and G.W. Barrett. 2006. *Cuterebra fontinella* parasitism on *Peromyscus leucopus* and *Ochrotomys nuttalli*. Southeastern Naturalist 5:157–164.

Kelly, D., and V.L. Sork. 2002. Mast seeding in perennial plants: Why, how, where? Annual Review of Ecology and Systematics 33:427–447.

Kitchings, J.T., and D.J. Levy. 1981. Habitat patterns in a small mammal community. Journal of Mammalogy 62:814–820.

Linzey, D.W. 1968. An ecological study of the golden mouse, *Ochrotomys nuttalli*, in the Great Smoky Mountains National Park. American Midland Naturalist 79:320–345.

Linzey, D.W., and R.L. Packard. 1977. *Ochrotomys nuttalli*. Mammalian Species 75:1–6.

Loeb, S.C. 1999. Responses of small mammals to coarse woody debris in a southeastern pine forest. Journal of Mammalogy 80:460–471.

Lomolino, M.V., and G.A. Smith. 2003. Prairie dog towns as islands: Applications of island biogeography and landscape ecology for conserving nonvolant terrestrial vertebrates. Global Ecology and Biogeography 12:275–286.

MacMahon, J.A. 1981. Successional processes: Comparisons among biomes with special reference to probable roles of and influences on animals. Pages 277–304 *in* D.C. West, H.H. Shugart, Jr., and D.B. Botkin, editors. Forest succession: Concepts and applications. Springer-Verlag, New York, New York.

Maser, C., J.M. Trappe, and R.A. Nussbaum. 1978. Fungal–small mammal interrelationshiops with emphasis on Oregon coniferous forests. Ecology 59:799–809.

McCarley, W.H. 1958. Ecology, behavior and population dynamics of *Peromyscus nuttalli* in eastern Texas. Texas Journal of Science 10:147–171.

McNaughton, S.J., R.W. Ruess, and S.W. Seagle. 1988. Large mammals and process dynamics in African ecosystems. BioScience 38:794–800.

Mengak, M.T., and D.C. Guynn, Jr. 2003. Small mammal microhabitat use on young loblolly pine regeneration areas. Forest Ecology and Management 173:309–317.

Mohan, J.E., L.H. Ziska, W.H. Schlesinger, R.B. Thomas, R.C. Sicher, K. George, and J.S. Clark. 2006. Biomass and toxicity responses of poison ivy (*Toxicodendron radicans*) to elevated atmospheric CO_2. Proceedings of the National Academy of Science 103:9086–9089.

Morzillo, M.T., G.A. Feldhamer, and M.C. Nicholson. 2003. Home range and nest use of the golden mouse (*Ochrotomys nuttalli*) in southern Illinois. Journal of Mammalogy 84:253–260.

Naiman, R.J., G. Pinay, C.A. Johnston, and J. Pastor. 1994. Beaver influences on the long-term biogeochemical characteristics of boreal forest drainage networks. Ecology 75:905–921.

Oliver, C.D., and B.C. Larson. 1990. Forest stand dynamics. McGraw-Hill, New York, New York.

O'Malley, M., J. Blesh, M. Williams, and G.W. Barrett. 2003. Food preferences and bioenergetics of the white-footed mouse (*Peromyscus leucopus*) and the golden mouse (*Ochrotomys nuttalli*). Georgia Journal of Science 61:233–237.

Orrock, J.L., and J.F. Pagels. 2002. Fungus consumption by the southern red-backed vole (*Clethrionomys gapperi*) in the southern Appalachians. American Midland Naturalist 147:413–418.

Orrock, J.L., D. Farley, and J.F. Pagels. 2003. Does fungus consumption by the woodland jumping mouse vary with habitat type or the abundance of other small mammals? Canadian Journal of Zoology 81:753–756.

Packard, R.L. 1969. Taxonomic review of the golden mouse, *Ochrotomys nuttalli*. Miscellaneous Publication 51, Pages 373–406. University of Kansas Museum of Natural History, Lawrence, Kansas.

Pastor, J., B. Dewey, and D.P. Christian. 1996. Carbon and nutrient mineralization and fungal spore composition of fecal pellets from voles in Minnesota. Ecography 19:52–61.

Peles, J.D., C.K. Williams, and G.W. Barrett. 1995. Bioenergetics of golden mice: The importance of food quality. American Midland Naturalist 133:373–376.

Perry, R.W., and R.E. Thill. 2005. Small-mammal responses to pine regeneration treatments in the Quachita Mountains of Arkansas and Oklahoma, USA. Forest Ecology and Management 219:81–94.

Phillips, O.L., R.V. Martinez, L. Arroyo, T.R. Baker, T. Killeen, S.L. Lewis, Y. Malhi, A.M. Mendoza, D. Neill, P.N. Vargas, M. Alexiades, C. Ceron, A. Di Fiore, T. Erwin, A. Jardim, W. Palacios, M. Saldias, and B. Vinceti. 2002. Increasing dominance of large lianas in Amazonian forests. Nature 418:770–774.

Rinker, H.B., M.D. Lowman, M.D. Hunter, T.D. Schowalter, and S.J. Fonte. 2001. Literature review: Canopy herbivory and soil ecology, the top-down impact of forest processes. Selbyana 22:225–231.

Schmid-Holmes, S., and L.C. Drickamer. 2001. Impact of forest patch characteristics on small mammal communities: A multivariate approach. Biological Conservation 99:293–305.

Schnurr, J.L., R.S. Ostfeld, and C.D. Canham. 2002. Direct and indirect effects of masting on rodent populations and tree seed survival. Oikos 96:402–410.

Seagle, S.W. 1983. Habitat availability and animal community characteristics. PhD dissertation, University of Tennessee, Knoxville, Tennessee.

Seagle, S.W. 1985a. Patterns of small mammal microhabitat utilization in cedar glade and deciduous forest habitats. Journal of Mammalogy 66:22–35.

Seagle, S.W. 1985b. Competition and coexistence of small mammals in an east Tennessee pine plantation. American Midland Naturalist 114:272–282.

Seagle, S.W. 2003. Can ungulates foraging in a multiple-use landscape alter forest nitrogen budgets? Oikos 103:230–234.

Sharitz, R.R., L.R. Boring, D.H. Van Lear, and J.E. Pinder, III. 1992. Integrating ecological concepts with natural resource management of southern forests. Ecological Applications 2:226–237.

Smith, M.H., J.B. Gentry, and J. Pinder. 1974. Annual fluctuations in small mammal populations in an eastern hardwood forest. Journal of Mammalogy 55:231–234.

Stueck, K.L., M.P. Farrell, and G.W. Barrett. 1977. Ecological energetics of the golden mouse based on three laboratory diets. Acta Theriologica 22:309–315.

Thomas, W.A. 1969. Accumulation and cycling of calcium by dogwood trees. Ecological Monographs 39:101–120.

Weltzin, J.F., R.T. Belote, and N.J. Sanders. 2003. Biological invaders in a greenhouse world: Will elevated CO_2 fuel plant invasions? Frontiers in Ecology and the Environment 1:146–153.

Wright, J.P., and C.G. Jones. 2006. The concept of organisms as ecosystem engineers ten years on: Progress, limitations, and challenges. BioScience 56:203–209.

6
Landscape Ecology of the Golden Mouse

Jerry O. Wolff and Gary W. Barrett

It is now abundantly clear that if we do not meet the challenge of landscape ecology, we can harbor little hope of stanching anthropogenic losses in Earth's biodiversity and hence of stemming the deterioration in the life support system of our own species. (Lidicker 1995:Preface vii)

Landscape ecology focuses on the development and dynamics of spatial heterogeneity, the influence of spatial heterogeneity on biotic and abiotic processes among ecosystems, and the management of spatial heterogeneity at the landscape scale. Landscapes are composed of three major elements: patches, corridors, and the landscape matrix. The role of habitat (patch) structure and composition on the dynamics of golden mice, *Ochrotomys nuttalli*, is relatively well known (Goodpaster and Hoffmeister 1954, Linzey 1968, McCarley 1958). Little is known, however, regarding metapopulation or source–sink dynamics, the role of corridors between and among landscape patches, or how the landscape mosaic influences dispersal, abundance, and reproductive success of this small mammal species. Our objective is to summarize the current state of knowledge regarding the landscape ecology of golden mice. We will first discuss golden mice regarding their response to landscape elements. Next, we postulate how landscape fragmentation impacts the social organization, dispersal, and colonization of golden mice. This chapter also illustrates the numerous unanswered questions and the research opportunities that exist at the landscape level.

Landscape Fragmentation

Much has been written during the past few decades describing the effects of habitat and landscape fragmentation on biotic diversity (e.g., Hilty et al. 2006). Connell (1979), for example, put forth the Intermediate Disturbance Hypothesis, which suggests that perturbations at intermediate levels of disturbance might maximize species diversity in a community, ecosystem, or landscape (Figure 6.1). In his book *The Fragmented Forest*, Harris (1984) described how island biogeography theory can be used in the preservation of biotic diversity. Harris also presented an excellent summary of home range sizes and linear-travel distances of mammals

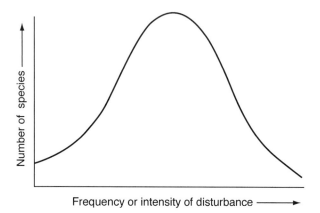

FIGURE 6.1. Diagram depicting the Intermediate Disturbance Hypothesis suggesting that the number of species in a community might be highest at intermediate levels of habitat disturbance. After Connell (1979).

ranging from deer mice (*Peromyscus maniculatus*) to grizzly bears (*Ursus arctos*) and explained how this parameter affects movements and colonization in a patchy environment. Patch size and quality, caused by landscape fragmentation, affect mammalian species differently based on interrelated factors such as body mass, home range size, and patterns of dispersal (see also Wolff 1999, 2003a). Because many species of small mammals select habitat with a dense understory, rodents such as golden mice likely benefit from frequent spatial-scale landscape fragmentation caused by disturbances such as timber harvest (Blus 1966, McCarley 1958), tornadoes (Loeb 1999), and fire (Masters et al. 1998).

These disturbances increase heavy undergrowth of vegetation that follows during the early stages of secondary succession. Thus, there likely exists an optimum level of disturbance for the maximal diversity of small mammal species.

Patch Quality

Golden mice have been described as a habitat specialist (Christopher and Barrett 2006, Dueser and Hallett 1980, Knuth and Barrett 1984, Linzey and Packard 1977, Seagle 1985). Golden mice prefer very dense mixed evergreen and deciduous canopies that include ample climbing undergrowth structures dominated by Japanese honeysuckle (*Lonicera japonica*), greenbrier (*Smilax glavca*), blackberry (*Rubus* sp.), and Chinese privet (*Ligustrum sinense*) (Christopher and Barrett 2006, Goodpaster and Hoffmeister 1954, Linzey 1968, Linzey and Packard 1977). Field studies have shown that golden mice were most frequently found 1.5 m above ground level, the level dominated by these understory shrubs (Christopher and Barrett 2006). A similar study in southern Illinois also reported that understory density 1.6–2.0-m-high provided optimum nesting habitat for golden mice

(Morzillo et al. 2003). Recently cut patches with an open canopy providing lush undergrowth appear to provide ideal habitat for golden mice (Schmid-Holmes and Drickamer 2001). Undergrowth species such as those described earlier frequently dominate habitat edges, which likewise provide ideal nesting habitat for golden mice (discussed later; see Figure 1.6 in Chapter 1 of this volume).

Changes in patch size caused by landscape fragmentation have been the subject of several small mammal investigations. For example, patch size has been shown to affect the movement patterns of cotton rats (*Sigmodon hispidus*), prairie voles (*Microtus ochrogaster*), and deer mice (Diffendorfer et al. 1995). The effects of corridors that connect landscape patches on home range sizes and interpatch movements of cotton rats, cotton mice (*Peromyscus gossypinus*), and old-field mice (*P. polionotus*) have similarly been investigated by Mabry and Barrett (2002). These investigations have shed light on the importance of home range size, patch connectivity, and patterns of movement on small mammal species at the landscape scale. Unfortunately, little is known regarding how patch size and patch connectivity specifically affect movement patterns of golden mice. The role of *landscape geometry* (the study of the shapes, patterns, and configurations of landscape patches) on small mammal population dynamics has been addressed to some extent by Harper et al. (1993) and Klee et al. (2004). See also Odum and Barrett (2005:399–404) for a detailed discussion of landscape geometry. Studies designed to address how landscape geometry and landscape architecture (patch stratification, "hard" versus "soft" edges, and three-dimensional use of habitat space) affect golden mice at the landscape scale provide a fertile field of investigation.

Landscape Corridors

Forested landscape corridors are important elements that enable animal dispersal among patches, reduce soil and wind erosion, allow gene transfer among patches, and provide habitat for nongame species such as golden mice. *Corridors* are classified into several basic types, including remnant, disturbance, planted, resource, and regenerated (Hilty et al. 2006). We focus our comments on only two corridor types: resource and disturbance. Rivers and streams are representative of resource corridors, including riparian vegetation (Figure 6.2). Rosenberg et al. (1997) provided examples of studies that quantify how the form, function, and efficacy of corridors affect small mammal movements between and among experimental landscape patches. The effects of corridor presence on population dynamics have been described for the meadow vole (*Microtus pennsylvanicus*; LaPolla and Barrett 1993). The interrelationship between home range size and interpatch movements have also been noted for cotton mice, old-field mice, and cotton rats (Mabry and Barrett 2002). Several investigators have reported golden mice in riparian habitats and corridors (Andrews 1963, Blus 1966, Christopher and Barrett 2006, Eads and Brown 1952, Handley 1948, Jennison et al. 2006, Pruett et al. 2002). However, no studies have been conducted to determine if golden mice use riparian corridors as paths of dispersal or movements related to metapopulation

FIGURE 6.2. A stream meandering through the countryside is representative of a riparian resource corridor—important for golden mice and other small mammals. Reprinted from Odum and Barrett 2005, with permission from Brooks/Cole, a division of Thomson Learning.

dynamics. Similarly, no studies have been conducted that attempt to quantify the effects of experimental or natural corridors on movements of golden mice.

A linear disturbance through the landscape matrix results in a disturbance corridor such as a power line through forest habitat (Figure 6.3). Disturbance corridors disrupt the natural homogeneous landscape matrix and provide important habitat for "opportunistic" species of plants and animals adapted to disturbance (Hilty et al. 2006). Golden mice are frequently found in disturbance habitats. In Madison County, Kentucky, for example, golden mice were collected by hand in an area disturbed by the placement of an electric power line (Knuth and Barrett 1984). Numerous nests of golden mice were also found along a power-line right-of-way in the Great Smoky Mountains in Tennessee (Linzey 1968). Following several tornadoes at the Savannah River Site in South Carolina, golden mice were captured only in unsalvaged plots (Loeb 1999), again indicative of a "disturbance" opportunistic

FIGURE 6.3. A power line cutting through a forest habitat is representative of a disturbance corridor. Like riparian corridors, they are important for small mammals. Reprinted from Odum and Barrett 2005, with permission from Brooks/ Cole, a division of Thomson Learning.

species. Golden mice also occur in areas disturbed by logging and road construction (Blus 1966, McCarley 1958). Shelterwood or selective logging tends to open the tree canopy and permit lush growth of understory species identified as important vegetation for golden mouse nests and reproduction (McCarley 1958). Thus, relatively detailed information suggests that golden mice are an "opportunistic species" adapted to and perhaps relying on either natural or anthropogenic disturbance.

Edge Habitat

The narrow zone or sharp demarcation of habitat between two community types (e.g., where a forest is directly adjacent to a crop field) is frequently termed *an edge*. An edge has long been considered to increase the abundance and diversity of plants and wildlife (Leopold 1933) and is known as *edge effect*. Species that use edges for purposes of reproduction and survival are frequently termed *edge species*. We suggest that the golden mouse be considered an edge species.

There exists a wealth of information describing the use of edge habitat by golden mice, including edge microhabitat (Seagle 1985), edge of swamps (Kitchings and Levy 1981), riparian streamsides (Miller et al. 2004), forest–farm edge (Heske 1995), edge of drainage ways (Andrews 1963), and the edge of power-line right-of-way corridors (Linzey 1968). Dense understory typically characterizes these edge habitats, which are frequently occupied by golden mice (Linzey 1968, McCarley 1958, Morzillo et al. 2003). The thick undergrowth vegetation serves as a substrate for nest construction, escape routes from the nest, and protection from predators (Klein and Layne 1978, Wagner et al. 2000). Because landscape fragmentation increases edge habitat, some level of fragmentation or disturbance might help to maintain populations of golden mice. However, in multiple-use management, consideration must be given to protecting the natural habitat for interior vertebrate species. For example, increased landscape fragmentation has benefited some edge species, but it has also had negative impacts on others, such as increasing the rate of brood parasitism on interior avian species by the brown-headed cowbird (*Molothrus ater*). The interaction of species along edge habitats can also affect community dynamics. For instance, selective grazing on tree seedlings by the grassland-dwelling meadow vole and competition with the forest-edge white-footed mouse (*Peromyscus leucopus*) can deter tree invasion in old-field ecosystems (Ostfeld et al. 1997). Similar studies investigating the relationship between golden mice and other small mammal species along edge habitats would increase understanding as to how vertebrate and plant communities are influenced by the dynamics of rodents at the landscape scale.

Landscape Matrix

Information regarding how the landscape matrix affects golden mouse abundance is severely lacking. For example, the longleaf pine (*Pinus palustris*) flatwoods in the Gulf Hammock area in northern Florida and Virginia pine (*P. virginiana*) in the Great Smoky Mountains National Park (Pearson 1954, Linzey 1968, respectively) are dominant matrix species. Do golden mice disperse and move across these matrices in search of another quality landscape patch? Cotton rats at the Savannah River Site in Aiken, South Carolina typically move from one experimental old-field patch to another across a forest matrix dominated by loblolly pine (*P. taeda*; Bowne et al. 1999, Mabry and Barrett 2002, Peles et al. 1999); however, it is not known if golden mice traverse this habitat as readily. The extent to which golden mice move across a landscape matrix either in search of another quality landscape patch or by just exploring it is not known. One study in which golden mice were radio-tracked demonstrated that males will readily cross a road to move from one nest to another; however, females appeared more reluctant to do so (Morzillo 2001). In this case, the mice may have just incorporated the roadway into their normal home ranges. To better predict the patterns of movement between and among landscape elements requires a further understanding of theoretical aspects of how social organization affects dispersal and colonization of fragmented landscapes.

Animals are not distributed randomly in time or space. Rather, their dispersion, movement, and distribution across a landscape are strongly dictated by their evolutionary history, dispersal ability, and social organization in conjunction with the distribution of suitable habitat (Wolff 1999, 2003a). We have very little data, and virtually no experiments, on the response of golden mice to fragmented forest landscapes. Therefore, in the following sections, we use evolutionary and behavioral theory to predict how individuals will be distributed across continuous and fragmented forest landscapes. First, we describe how the social organization of a species dictates its use of space and impacts its pattern of dispersal. Second, we use behavioral and ecological theory to predict how golden mice will disperse across matrix habitat, colonize fragmented landscapes, and exist in a metapopulation. Our perspective on how social organization of golden mice affects their dispersal and dispersion in the landscape mosaic is illustrated in Figure 6.4.

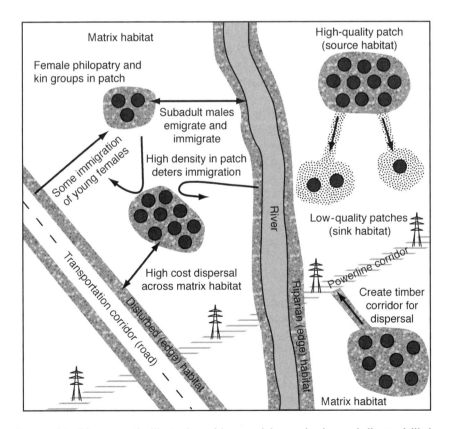

FIGURE 6.4. Diagrammatic illustration of how social organization and dispersal likely occur among high- and low-quality patches, natural and constructed corridors, in the landscape matrix. Circles in patches represent occupied territories.

Territoriality and the Dispersion of Exclusive Breeding Space

The main factor that determines how animals use space and move across the landscape is whether they follow an Ideal Free Distribution or an Ideal Despotic Distribution (Fretwell and Lucas 1970). Species in which either males or females actively defend space against conspecifics (i.e., are territorial) are distributed despotically. Conversely, free distribution occurs in species that do not defend territories. Thus, the presence of resident individuals affects how dispersing or immigrating individuals will move across the landscape and colonize occupied versus vacant habitat (Wolff 1997, 2003a). In rodents, breeding adult females despotically defend exclusive space with respect to unrelated adult females. This pattern of territorial defense, in turn, influences dispersal, juvenile recruitment, use of source and sink habitats, and the potential for colonization (Wolff 1997, 1999, 2003a, 2003b). Aggression among breeding females is high, apparently to protect their offspring from intruding females that commit infanticide as a means of acquiring breeding space for themselves (Wolff and Peterson 1998). Consequently, the size of the breeding population, which, in turn, affects the rate of population growth and rate of immigration and emigration, should be dependent to some extent on the number of exclusive breeding sites available to females.

The social organization of golden mice has not been studied in detail. From what we have learned indirectly from field observations and can predict from behavioral theory, however, breeding females occupy exclusive space with respect to unrelated females and defend a space around their nest site from other adult females. Although golden mice are reportedly quite docile (Dunaway 1955, Goodpaster and Hoffmeister 1954, Linzey and Packard 1977), reproductive females are likely quite aggressive toward intruders that potentially might harm their offspring. The main factor that affects recruitment of juveniles in rodents is the density of adult females in the population (Krebs et al. 2007, Wolff 2003b, Wolff et al. 2002). Juvenile recruitment typically is inversely related to the number of adult females in a population. Thus, the dispersion of females defending exclusive space against intruding females likely affects the size of the breeding population.

Dispersal and Philopatry

Dispersal in golden mice has not been investigated directly but is likely similar to that of *Peromyscus* and most other rodents (Nunes 2007). The typical dispersal pattern for most species of rodents is for juvenile males to emigrate from the maternal site shortly after weaning and for daughters to establish a residence near—often adjacent to—the natal site (Greenwood 1980, Pusey 1987, Wolff 1993, 1994). This pattern of philopatry apparently occurs in golden mice because most females appear to stay near their natal site rather than disperse. Philopatry is relatively common in female rodents such as golden mice because of (1) the benefits of nesting near the maternal site, which is a proven area for successfully rearing young, (2) nepotistic behavior of sharing adjacent space with female relatives, and (3) costs and uncertainties associated with dispersing to a new landscape patch.

We expect juvenile males will disperse from the natal site to seek unrelated mates and to avoid inbreeding with female relatives who remain in the natal site (Wolff 1994). Daughters typically will be relatively philopatric, nesting near their maternal site. Males and other family members might nest communally during the nonbreeding season such as late autumn or winter months (Barbour 1942, Dietz and Barrett 1992, Goodpaster and Hoffmeister 1954); however, dispersal should occur prior to breeding. The dispersal distance for mammals is highly variable. The dispersal distance averages about 10 home range diameters (Van Vuren 1998, Wolff 1999), but the actual distance is likely a function of availability of mates, suitable breeding habitat, and geometry of the landscape. The nesting behavior of the golden mouse is discussed in more detail in Chapter 9 of this volume.

Emigration appears to be an adaptive tactic for an individual to seek better habitat and/or better mating opportunities (Wolff 1993, 1994). However, the rate at which individuals disperse from their natal site is, in part, a function of the ease with which they can immigrate into new space. At low densities and when suitable habitat is available for colonization, individuals can emigrate and immigrate relatively easily without deterrence by territorial residents. However, at high densities when all suitable habitat is occupied, territoriality can serve as a barrier to animal movement (Figure 6.4). In contiguous habitat, a high-density population of golden mice can deter immigration and emigration. However, individuals on the edge of a patch, a preferred habitat for golden mice, should be able to disperse without any impediment. Unfortunately, those individuals that live on the edge of a patch are likely confronted with sink or matrix habitat. Thus, emigration would have a greater energetic cost and risk of predation than it would across more suitable habitat or along corridors.

Once a male or female golden mouse establishes residence, there is little evidence that they emigrate from their home site. Golden mice appear to be more sedentary than many other species of small mammals (Komarek 1939). In fact, one individual that was livetrapped over a span of 1.5 years was last captured within 30 m of its initial point of release (Pearson 1953). Not only is there little evidence of adult dispersal, but the home range size is small as well. For example, the mean home range size for golden mice, based on live trapping data, is ~0.05–0.57 ha (McCarley 1958, Redman and Sealander 1958). Linzey (1968) reported an average home range size of 0.26 ha for males and 0.24 ha for females. The mean home range size as revealed by radiotelemetry in a riparian forested landscape peninsula was ~0.90 ha for males and ~0.50 ha for females (Pruett et al. 2002), and in a southern Illinois forest, it was 0.31–4.64 ha for males and 0.09–0.90 ha for females (Morzillo 2001). Thus, a small home range size and a sedentary nature suggest that adults do not readily traverse matrix habitat and are likely not good colonists. If dispersal and colonization of new habitat patches are going to occur by golden mice, it will likely be by subadults.

Sex-Biased Dispersal and Colonization

One of the main factors that determines the dispersion of a species across a landscape mosaic is its ability to disperse and colonize new habitats. For a species to be a good colonist, females must disperse, become successful immigrants, and reproduce. Considering that females of most mammal species are relatively philopatric and remain in or near their natal site, mammals, in general, are not good colonists. Dispersal distances obviously differ for various species of mammals and correlate allometrically with body mass (Van Vuren 1998, Wolff 1999). Maximum dispersal distances for females are often comparable to those of males; however, the proportion of females dispersing is low and the median dispersal distances are generally short. For a mammal the size of the golden mouse (20–25 g), anticipated mean dispersal distance would be about 300–500 m. In that golden mice have rather specific habitat requirements (described in Chapters 1 and 9 of this volume), dispersal distances might vary considerably depending on finding suitable habitat. Thus, the probability of colonizing and establishing a breeding population at new sites or in distant patches often is lower than would be predicted based on an estimated dispersal distance for the species.

Source–Sink Dynamics

Source and sink habitats probably occur for all species, but they have the greatest impact on terrestrial species that exhibit a despotic distribution (Pulliam 1988, Wolff 1999). Because adult female golden mice occupy exclusive space during the breeding season, once suitable habitat becomes occupied, subordinate or dispersing individuals are relegated to sink habitats. Source and sink habitats for golden mice are a function of the landscape mosaic and dispersion of suitable brushy edge and patch habitats interspersed among matrix habitat and their connectivity by corridors (Figure 6.4). We envision corridors as very important for this species to survive the dispersal phase in searching for a suitable patch— patches that are highly fragmented and often widely dispersed.

Mabry and Barrett (2002) described the effects and importance of landscape corridors on home range sizes and interpatch movements of small mammals at the experimental landscape scale. Golden mice live and nest in edge and riparian habitats (Christopher and Barrett 2006, Klee et al. 2004). However, because golden mice are habitat specialists (Christopher and Barrett 2006), they might have a more difficult time crossing a diverse matrix or sink habitat than would a habitat generalist such as *P. leucopus* (Klee et al. 2004, Wolff 1999, 2003a). It is possible that edges serve as corridors for source–sink dynamics (Figure 6.4), but current studies have not been designed to monitor linear- and long-distance movements of golden mice. Riparian (stream) resource and disturbance corridors could similarly serve as dispersal corridors within which golden mice reside (Figure 6.4). The relevancy of source–sink dynamics to golden mice awaits further investigation, especially regarding the interspersion and relative frequency distribution of patches, corridors, and matrix habitat.

Summary and Future Directions

Golden mice are an excellent model species for studies on metapopulation dynamics because of their habitat specialization and persistence in fragmented and disturbed landscapes. Golden mice live in a patchy environment and, thus, must possess traits that allow them to disperse, follow corridors, and move through a matrix habitat to colonize often distant and unpredictable habitat patches. The social organization appears to provide for cooperation among kin members for communal nesting and sharing habitat when dispersal is difficult (Dietz and Barrett 1992) while providing opportunities for emigration to promote outbreeding. Although few studies have experimentally or theoretically addressed the role of social behavior in regulating the dispersion, spacing, and interdemic movements of golden mice, related studies on other rodents provide the basis for developing *a priori* predictions for how this species likely functions in fragmented landscapes. Future studies should use behavioral and ecological theory to develop and test hypotheses to determine the proximate causation and ultimate benefits for dispersal and colonization that result in the pattern of distribution of golden mice in fragmented landscapes.

Literature Cited

Andrews, R.D. 1963. The golden mouse in southern Illinois. Chicago Academy of Sciences Natural History Miscellanea 179:1–3.

Barbour, R.W. 1942. Nests and habitat of the golden mouse in eastern Kentucky. Journal of Mammalogy 23:90–91.

Blus, L.J. 1966. Some aspects of golden mouse ecology in southern Illinois. Transactions of Illinois State Academy of Science 59:34–41.

Bowne, D.R., J.D. Peles, and G.W. Barrett. 1999. Effects of landscape spatial structure on movement patterns of the hispid cotton rat (*Sigmodon hispidus*). Landscape Ecology 14:53–65.

Christopher, C.C., and G.W. Barrett. 2006. Coexistence of white-footed mice (*Peromyscus leucopus*) and golden mice (*Ochrotomys nuttalli*) in a southeastern forest. Journal of Mammalogy 87:102–107.

Connell, J.H. 1979. Tropical rainforests and coral reefs as open nonequilibrium systems. Pages 141–163 *in* R.M. Anderson, B.D. Turner, and L.R. Taylor, editors. Population dynamics. Blackwell, Oxford, United Kingdom.

Dietz, B.A., and G.W. Barrett. 1992. Nesting behavior of *Ochrotomys nuttalli* under experimental conditions. Journal of Mammalogy 73:577–581.

Diffendorfer, J.E., M.S. Gaines, and R.D. Holt. 1995. Habitat fragmentation and movements of three small mammals (*Sigmodon, Microtus,* and *Peromyscus*). Ecology 76:827–839.

Dueser, R.D., and J.G. Hallett. 1980. Competition and habitat selection in a forest-floor small mammal fauna. Oikos 35:293–297.

Dunaway, P.B. 1955. Late fall home ranges of three golden mice, *Peromyscus nuttalli.* Journal of Mammalogy 36:297–298.

Eads, J.H., and J.S. Brown. 1952. Studies of the golden mouse, *Peromyscus nuttalli aureolus,* in Alabama. Journal of the Alabama Academy of Science 25:25–26.

Fretwell, S.D., and H.L. Lucas. 1970. On territorial behaviour and other factors influencing habitat distribution in birds. Theoretical Development, Vol. 1. Acta Biotheoretica 19:16–36.

Goodpaster, W.W., and D.F. Hoffmeister. 1954. Life history of the golden mouse, *Peromyscus nuttalli,* in Kentucky. Journal of Mammalogy 35:16–27.

Greenwood, P.J. 1980. Mating systems, philopatry and dispersal in birds and mammals. Animal Behaviour 28:1140–1162.

Handley, C.O., Jr. 1948. Habitat of the golden mouse in Virginia. Journal of Mammalogy 29: 298–299.

Harper, S.J., E.R. Bollinger, and G.W. Barrett. 1993. Effects of habitat patch shape on population dynamics of meadow voles (*Microtus pennsylvanicus*). Journal of Mammalogy 74:1045–1055.

Harris, L.D. 1984. The fragmented forest. University of Chicago Press, Chicago, Illinois.

Heske, E.J. 1995. Mammalian abundances on forest-farm edges versus forest interiors in southern Illinois: Is there an edge effect? Journal of Mammalogy 76:562–568.

Hilty, J.A., W.Z. Lidicker, Jr., and A.M. Merenlender. 2006. Corridor ecology. Island Press, Washington DC.

Jennison, C.A., L.R. Rodas, and G.W. Barrett. 2006. *Cuterebra fontinella* parasitism on *Peromyscus leucopus* and *Ochrotomys nuttalli.* Southeastern Naturalist 5:157–164.

Kitchings, J.T., and D.J. Levy. 1981. Habitat pattern in a small mammal community. Journal of Mammalogy 62:814–820.

Klee, R.V., A.C. Mahoney, C.C. Christopher, and G.W. Barrett. 2004. Riverine peninsulas: An experimental approach to homing in white-footed mice (*Peromyscus leucopus*). American Midland Naturalist 151:408–413.

Klein, H.G., and J.N. Layne. 1978. Nesting behavior in four species of mice. Journal of Mammalogy 59:103–108.

Knuth, B.A., and G.W. Barrett. 1984. A comparative study of resource partitioning between *Ochrotomys nuttalli* and *Peromyscus leucopus.* Journal of Mammalogy 65:576–583.

Komarek, E.V. 1939. A progress report on southeastern mammal studies. Journal of Mammalogy 20: 292–299.

Krebs, C.J., X. Lambin, and J.O. Wolff. 2007. Social behavior and self regulation in Murid rodents. Pages 173–181 *in* J.O. Wolff and P.W. Sherman, editors. Rodent societies: An ecological and evolutionary perspective. University of Chicago Press, Chicago, Illinois.

LaPolla, V.N., and G.W. Barrett. 1993. Effects of corridor width and presence on the population dynamics of the meadow vole (*Microtus pennsylvanicus*). Landscape Ecology 8:25–37.

Leopold, A. 1933. Game management. Charles Scribner and Sons, New York, New York.

Lidicker, W.Z., Jr. 1995. Page 251 *in* Landscape approaches in mammalian ecology and conservation. University of Minnesota Press, Minneapolis, Minnesota.

Linzey, D.W. 1968. An ecological study of the golden mouse, *Ochrotomys nuttalli*, in the Great Smoky Mountains National Park. American Midland Naturalist 79:320–345.

Linzey, D.W., and R.L. Packard. 1977. *Ochrotomys nuttalli*. Mammalian Species 75:1–6.

Loeb, S.C. 1999. Responses of small mammals to course woody debris in a southeastern pine forest. Journal of Mammalogy 80:460–471.

Mabry, K.E., and G.W. Barrett. 2002. Effects of corridors on home range sizes and interpatch movements of three small mammal species. Landscape Ecology 17:629–636.

Masters, R.E., R.L. Lochmiller, S.T. McMurry, and G.A. Buhenhoffer. 1998. Small mammal response to pine-grassland reforestation for red-cockaded woodpeckers. Wildlife Society Bulletin 26:148–158.

McCarley, W.H. 1958. Ecology, behavior and population dynamics of *Peromyscus nuttalli* in eastern Texas. Texas Journal of Science 10:147–171.

Miller, D.A., R.E. Thill, M.A. Melchoirs, T.B. Wigley, and P.A. Tappea. 2004. Small mammal communities of streamside management zones in intensively managed pine forests of Arkansas. Forest Ecology and Management 203:381–393.

Morzillo, A.T. 2001. Nest and habitat use by the golden mouse (*Ochrotomys nuttalli*) in southern Illinois. MS thesis, Southern Illinois University, Carbondale, Illinois.

Morzillo, A.T., G.A. Feldhamer, and M.C. Nicholson. 2003. Home range and nest use of the golden mouse (*Ochrotomys nuttalli*) in southern Illinois. Journal of Mammalogy 84:553–560.

Nunes, S. 2007. Dispersal. Pages 150–162 *in* J.O. Wolff and P.W. Sherman, editors. Rodent societies: An ecological and evolutionary perspective. University of Chicago Press, Chicago, Illinois.

Odum, E.P., and G.W. Barrett. 2005. Fundamentals of ecology, 5th ed. Thomson Brooks/Cole, Belmont, California.

Ostfeld, R.S., R.H. Manson, and C.D. Canham. 1997. Effects of rodents on survival of tree seeds and seedlings invading old fields. Ecology 78:1531–1542.

Pearson, P.G. 1953. A field study of *Peromyscus* populations in Gulf Hammock, Florida. Ecology 34:199–207.

Pearson, P.G. 1954. Mammals of Gulf Hammock, Levy County, Florida. American Midland Naturalist 51:468–480.

Peles, J.D., D.R. Bowne, and G.W. Barrett. 1999. Influence of landscape structure on movement patterns of small mammals. Pages 41–62 *in* G.W. Barrett and J.D. Peles, editors. Landscape ecology of small mammals. Springer-Verlag, New York, New York.

Pruett, A.L., C.C. Christopher, and G.W. Barrett. 2002. Effects of a riparian peninsula on mean home range size of the golden mouse (*Ochrotomys nuttalli*) and the white-footed mouse (*Peromyscus leucopus*). Georgia Journal of Science 60:201–208.

Pulliam, H.R. 1988. Sources, sinks, and population regulation. American Naturalist 132:652–661.

Pusey, A.E. 1987. Sex-biased dispersal and inbreeding avoidance in birds and mammals. Trends in Ecology and Evolution 2:295–299.

Redman, J.P., and J.A. Sealander. 1958. Home ranges of deer mice in southern Arkansas. Journal of Mammalogy 39:390–395.

Rosenberg, D.K., B.R. Noon, and E.C. Meslow. 1997. Biological corridors: Form, function and efficacy. BioScience 47:677–687.

Schmid-Holmes, S., and L.C. Drickamer. 2001. Impact of forest patch characteristics on small mammal communities: A multivariate approach. Biological Conservation 99:293–305.

Seagle, S.W. 1985. Competition and coexistence of small mammals in an east Tennessee pine plantation. American Midland Naturalist 114:272–282.

Van Vuren, D. 1998. Mammalian dispersal and reserve design. Pages 363–393 *in* T.M. Caro, editor. Behavioral ecology and conservation biology. Oxford University Press, Oxford, United Kingdom.

Wagner, D.M., G.A. Feldhamer, and J.A. Newman. 2000. Microhabitat selection by golden mice (*Ochrotomys nuttalli*) at arboreal nest sites. American Midland Naturalist 144:220–225.

Wolff, J.O. 1993. What is the role of adults in mammalian juvenile dispersal? Oikos 68:173–176.

Wolff, J.O. 1994. More on juvenile dispersal in mammals. Oikos 71:349–352.

Wolff, J.O. 1997. Population regulation in mammals: An evolutionary perspective. Journal of Animal Ecology 66:1–13.

Wolff, J.O. 1999. Behavioral model systems. Pages 11–40 *in* G.W. Barrett and J.D. Peles, editors. Landscape ecology of small mammals. Springer-Verlag, New York, New York.

Wolff, J.O. 2003a. An evolutionary and behavioral perspective on dispersal and colonization of mammals in fragmented landscapes. Pages 614–630 *in* C.J. Zabel and R.G. Anthony, editors. Mammalian community dynamics: Management and conservation in the coniferous forests of western North America. Cambridge University Press, Cambridge, United Kingdom.

Wolff, J.O. 2003b. Density-dependence and the socioecology of space use in rodents. Pages 124–130 *in* G.R. Singleton, L.A. Hinds, C.J. Krebs, and D.M. Spratt, editors. Rats, mice and people: Rodent biology and management. Australian Centre for International Agricultural Research, Canberra, Australia Monograph 96.

Wolff, J.O., and J.A. Peterson. 1998. An offspring-defense hypothesis for territoriality in female mammals. Ethology, Ecology and Evolution 10:227–239.

Wolff, J.O., W. D. Edge, and G. Wang, 2002. Effects of adult sex ratios on recruitment of juvenile gray-tailed voles. Journal of Mammalogy 83:947–956.

Section 3
Transcending Processes

7

Relative Abundance and Conservation: Is the Golden Mouse a Rare Species?

GEORGE A. FELDHAMER AND ANITA T. MORZILLO

The golden mouse, although widely distributed, is nowhere abundant (Hall and Kelson 1959:657)

Most mammalian species in North America are neither widespread nor abundant, and as noted by Kunin and Gaston (1993:298), ". . . most of the world's species are rare in some sense of the word." Rarity of species—and their enhanced potential for extinction—is a fundamental concept in conservation biology. Nonetheless, there is little consensus as to the meaning of the term "rare" in the biological literature (cf. Gaston 1994a, 1997, Munton 1987). Although implicitly understood to be "not common," rare has been defined in a ". . . variety of different ways and at a range of spatial scales" such that ". . . studies are seldom directly comparable in any but the broadest qualitative sense" (Gaston and Kunin 1997:12).

Resource managers working with rare species that might be identified as threatened or endangered at the state or national level face many ecological and methodological uncertainties. A species might actually be rare (very low abundance). Conversely, it might only appear to be rare because it is highly elusive, spatially clustered throughout its range (McDonald 2004), or exhibits temporal variation in abundance. Numerous sampling methods have been proposed to identify these possibilities (Thompson 2004). Regardless of definitions or sampling approaches, however, there are several ecological factors that determine whether a species is common or rare. In this chapter, we consider ecological factors affecting where the golden mouse occurs on a continuum from common to rare and how rarity might play a role in its current conservation status throughout its range. Rabinowitz (1981) and Rabinowitz et al. (1986) discussed species rarity in terms of three factors: geographic range, local population abundance, and diversity (quality) of habitats occupied. A matrix for rarity or commonness of a species emerges (Table 7.1) when the two end points of a continuum for each of these factors are considered for mammals (Yu and Dobson 2000). If a species falls in the small/low/narrow range for each factor, obviously it would be expected to be rare—category D in Table 7.1. So, where does the golden mouse occur in this matrix of geographic range, local abundance, and diversity of habitat types occupied?

TABLE 7.1. A matrix of three factors related to abundance or rarity of a mammalian species.

		Geographic range			
		Large		Small	
Local population abundance		High	Low	High	Low
Habitats occupied	Broad	A	B	B	C
	Narrow	B	C	C	D

Source: Adapted from Rabinowitz (1981) and Yu and Dobson (2000).
Note: We designate the resulting categories as (A) abundant, (B) fairly common, (C) fairly uncommon, or (D) rare.

Geographic Range

It might be somewhat surprising that the geographic range of most species of North American mammals is fairly limited. Pagel et al. (1991) examined the ranges of 679 North American species as determined from maps in Hall (1981). They found the median geographic range of these mammalian species was only about 1 percent of the total area of North America; only about 14 species had ranges >50 percent of the area of North America. Pimm and Jenkins (2005) put these data in perspective when they stated that one in six species of North American mammals has a range smaller than the state of Connecticut. Most have ranges smaller than the states of California, Oregon, and Washington combined. Also, the size of geographic range shows well-established latitudinal and longitudinal trends—ranges in North America tend to be larger in the north and east (McCoy and Connor 1980, Rapoport 1982, Stevens 1989). As might be expected, rodents were among several orders whose species generally have relatively restricted ranges.

Although a variety of methods has been used to define ranges (Gaston 1994b), identifying the geographic range of a species is not always straightforward. A range map of a species is delimited by latitudinal and longitudinal boundaries. However, this "extent of occurrence" will always be larger than the actual "area of occupancy" of a species within its range (Gaston 1997). Determining either area becomes more problematic when working with small, cryptic, and uncommon species. Nonetheless, there is a fairly well-documented positive relationship between geographic range and body size in mammals—larger species have larger ranges and smaller species tend to have smaller ranges. Gaston and Blackburn (1996) agreed that large species have larger ranges but suggested that small species exhibit a variety of range sizes from small to large. The golden mouse is a prime example of a small species with a relatively extensive geographic range (see Figure 2.1 Chapter 2 of this volume). The extent of occurrence of golden mice is well above the average for most North American mammals, especially for a small rodent at a southern latitude.

Local Abundance

As a general rule, mammalian species with smaller ranges are locally scarce, whereas those with larger ranges are locally common. The golden mouse does not fit this generalization, however. Despite enjoying a relatively large geographic range, local populations of golden mice are rarely abundant relative to white-footed mice (*Peromyscus leucopus*) or other sympatric rodents. Many investigators attest to relatively low population densities of golden mice (see Rose Chapter 3 of this volume). For example, density averaged about 4 residents/ha on four of five plots in southern Illinois, where Blus (1966a:341) reported, "On all plots, there were months of intensive trapping when no [golden] mice were taken." Howell (1954) in Tennessee, McCarley (1958) in Texas, Shadowen (1963) in Louisiana, and Feldhamer and Maycroft (1992) in Illinois reported similar low densities for the golden mouse. Also likely reflecting low population densities, substantially fewer golden mice were trapped relative to white-footed mice in studies by Dueser and Hallett (1980), Dueser and Shugart (1978), Feldhamer and Paine (1987), and Kitchings and Levy (1981). In forested habitat in Florida, Pearson (1953) caught almost four times more cotton mice (*P. gossypinus*) than golden mice. Pruett et al. (2002) livetrapped 61 golden mice on a 0.2-ha study site. Nonetheless, they caught 146 white-footed mice on the site during the same period. Contrary to this general trend, Miller et al. (2004) took more golden mice than any other small mammal species along riparian forest zones in Arkansas, although captures per trap night were very low. Like other species, the local abundance of the golden mouse exhibits temporal variation, often with predictable seasonal fluctuations related to reproductive pulses (Linzey 1968a). Nonetheless, throughout its range *Ochrotomys nuttalli* is usually scarce relative to sympatric small mammals.

Diversity of Habitat Types

Pagel et al. (1991) suggested that North American mammalian species in more southerly latitudes occur in fewer distinct habitat types. Locally, as noted in Chapter 2 of this volume, the golden mouse might occupy deciduous hardwood forests, coniferous forests, old fields, swampy lowlands, riparian bottomlands, and xeric wooded uplands. In more general terms, however, these classifications can all be considered forested habitat and the golden mouse can be considered a fairly restricted habitat specialist. Many investigators have considered the golden mouse to be a habitat specialist (Dueser and Hallett 1980, Knuth and Barrett 1984, Seagle 1985). Using climate and vegetation criteria for North America, Aldrich (1963) identified 18 major habitat types. Despite their fairly extensive geographic range, golden mice occur in only two of these—Eastern Deciduous Forest and Southern Evergreen; these habitats do make up a significant portion of the eastern United States.

Additional Considerations

Several other factors affect the abundance or rarity of a species in conjunction with geographic distribution, local abundance, and the diversity of habitats occupied. These include trophic level, body size, competitive ability, niche specialization, reproductive potential, dispersal ability, and genetic polymorphism. Many of these factors can be difficult to assess, with studies constrained and confounded by interactions among the various factors. As a result, conclusions are often accompanied with ". . . various caveats and probably many exceptions . . . often with substantial unexplained variance and only moderate explanatory power" (Gaston and Kunin 1997:22). Thus, sometimes only broad generalizations are possible about how these and other factors might affect abundant and rare species.

Trophic Level and Body Size

Species at higher trophic levels generally have lower abundance; mammalian carnivores (secondary consumers) are less abundant than herbivores (primary consumers) of similar body size. Conversely, geographic and home range sizes are generally larger for mammals at higher trophic levels. Given that most upper-level carnivores are large, body size is additional interacting factor when considering the affect of trophic level on abundance. There is a general inverse relationship between abundance and body size of animals. For example, Damuth (1981) found that the relationship between population density (number/km^2) and body weight (g) of 307 primary mammalian consumers was Density = 4.23 (Weight$^{-0.75}$). There has been much debate over the validity and characteristics of this overall relationship—depending on what data are analyzed, how they are analyzed, and at what scale (cf. Congreave 1993, Currie 1993, Damuth 1993, Lawton 1989). Regardless, investigators since Elton (1927) have agreed on the generalization that large-bodied animals are less common than small-bodied ones. Obviously, however, numerous interacting factors affect abundance (Figure 7.1); being small does not guarantee being abundant, because as noted, most species are rare.

Interspecific Competition and Niche Width

Competitive ability certainly might be expected to play a role in the abundance, dispersal, and geographic range of species. It is intuitively compelling to believe that rare species are poor competitors, but this might not always be the case. Past studies of interspecific competition in the golden mouse are equivocal. Given the range, habitat, and life-history characteristics of golden mice, the most common potential competitors often are white-footed mice and cotton mice. As noted previously, *Peromyscus* most often outnumber golden mice locally. The extent to which this disparity in numbers results from interspecific competitive interactions is unresolved and might vary locally. Dueser and Hallett (1980), for example, concluded that competition was occurring between

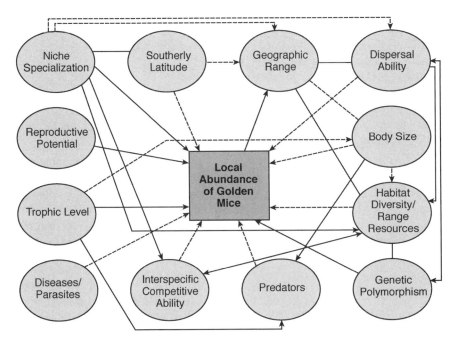

FIGURE 7.1. A conceptual model of the interactions among various ecological factors that tend to either increase (solid arrows) or decrease (dashed arrows) the local population abundance of golden mice.

golden mice and white-footed mice on mesic forest stands in east Tennessee. They considered the white-footed mouse "a poorly competitive habitat generalist" and the golden mouse "a strongly competitive habitat specialist." Seagle (1985) found the niche space of golden mice increased upon removal of white-footed mice, apparently resulting from competitive release. However, a similar removal study in southern Illinois (Corgiat 1996) found no change in space use by golden mice. Removal of *P. leucopus* by Christopher and Barrett (2006) had no affect on the abundance of golden mice, although *O. nuttalli* were taken more often in traps on the ground when *P. leucopus* were absent, indicating a shift in use of habitat space. Feldhamer and Maycroft (1992) suggested that either mutual avoidance or selection for different microhabitats by these two species caused an inverse relationship in densities on pine forest sites throughout southern Illinois. Also, there were significantly fewer trap sites where both species were captured than would be expected by chance. However, Christopher and Barrett (2007) took golden mice and white-footed mice simultaneously in the same traps. In a study on the same site, Pruett et al. (2002) concluded that mutualism (protocooperation) rather than competition might be enhancing coexistence of these two species.

Reproduction

The reproductive potential of a species obviously has a significant impact on population abundance and regulation. Site-specific factors are of major importance in reproductive potential, but there might be geographic considerations as well. In North America, Glazier (1980) found smaller mean litter sizes and less maternal energy expenditure per litter in geographically restricted species of *Peromyscus* compared to more widely distributed species. Conversely, there was no relationship between abundance and fecundity of 16 nonvolant mammalian species in rain forests of Australia (Laurance 1991). There is a well-defined latitudinal gradient with litter sizes (Lord 1960), in that larger litters occur in more northerly parts of a range. Golden mice appear to follow this trend (Blus 1966b; see Chapter 3 of this volume). Regardless, the reproductive characteristics of golden mice—in terms of such factors as age at first breeding, mean litter size, and number of litters per year—and the resulting impacts on population dynamics appear to be similar (Linzey and Packard 1977) to commonly sympatric species such as white-footed mice and cotton mice.

Dispersal

Dispersal—defined as individuals leaving their natal area and not returning—is a major component of social behavior, population regulation, and abundance. Dispersal also is a critical component of population genetics (Barton 1992, Lidicker 1975, Lidicker and Patton 1987). Dispersal usually commences postweaning, although certain species disperse later in life. Other species might never disperse but remain in the area where they were born. A species with high dispersal ability might be better able to find high-quality habitat patches, develop a larger geographic range, and reach higher population abundance than a species that is a poor disperser. In mammals, males usually are more prone to disperse than females, and juveniles more than adults. There are a variety of hypotheses regarding the reasons for dispersal in rodents in terms of relative individual benefits and costs, when individuals leave their natal area, and how far they move (cf. Ebensperger 2001, Greenwood 1980, Solomon 2003, Stenseth and Lidicker 1992). For example, benefits to dispersal would accrue if there were greater probabilities of finding better food resources, increased access to nesting sites or mates, reduced inbreeding, or fewer potential predators or competitors away from the natal area. Conversely, associated costs of dispersal could include greater energy expenditure of movement and finding mates, decreased time to fully mature, and unfamiliarity with a new area. These and other factors are all synergistic; a mosaic of ecological and environmental factors (Howard 1960, Lidicker and Stenseth 1992) interacts to affect the degree of dispersal among individuals in local populations and, consequently, the abundance or rarity of a species.

Most studies of dispersal in *Peromyscus* have involved white-footed mice or deer mice (*P. maniculatus*) (Wolff 1989). For example, Wolff et al. (1988) found that dispersal of young male *P. leucopus* was primarily related to reduced inbreeding, as did Nadeau et al. (1981) as well as Krohne et al. (1984).

Although early investigators characterized golden mice as docile and fairly sedentary with limited home ranges (Komarek 1939, Linzey 1968b, McCarley 1959, Shadowen 1963), individuals might move greater distances on a local level than previously believed (Morzillo et al. 2003). Dispersal patterns on a landscape level have not been investigated in golden mice, but dispersal might be limited by small body size, habitat specialization, or patch connectivity. Also, species with slow developmental rates are predicted to delay dispersal, but golden mice mature rapidly—at least relative to white-footed mice (Layne 1960, Linzey and Linzey 1967). There are no data regarding dispersal and how potential benefits and costs apply to golden mice populations; patterns might be similar to those in *Peromyscus*. See Chapter 6 of this volume for additional details regarding possible dispersal, colonization, and social organization at the landscape level.

Genetic Polymorphism

Species that have large populations often have greater genetic diversity than do rare species because the latter have smaller populations that are more prone to genetic drift. Analyses of both macrogeographic and microgeographic genetic variation in a variety of species of *Peromyscus* have played a major role in studies of mammalian population genetics (Kaufman and Kaufman 1989). These studies have shown significant genetic variation at all levels (Durish et al. 2004). We are unaware, however, of any comparable data for golden mice.

As noted, all the factors that impact on relative abundance of the golden mouse are interactive (Figure 7.1); examining single factors might not be only of limited value but also might be misleading. Additionally, the interactive factors affecting abundance are dynamic spatially and temporally. Thus, a species might be rare at the landscape level yet be fairly common at some local sites, at least during certain times of the year. Overall, we suggest that the golden mouse exhibits what Schoener (1987) called "suffusive rarity," in that the species is usually uncommon throughout its moderately extensive geographic range. Given the difficulty of assessing many of the ecological factors affecting golden mouse abundance, it might be expected that its conservation status at the state level is variable within the geographic range. Some states might lack the data necessary for informed management decisions.

Conservation Status

The conservation status of the golden mouse varies across its geographic range because of two main factors: (1) the level of assessment (either national or state) and (2) whether states are on the periphery of its geographic range. At the national level, the World Conservation Union Red Data List ([IUCN] International Union for Conservation of Nature and Natural Resources, Gland, Switzerland) considers *O. nuttalli* a species of "least concern" (Baillie 1996). Populations are considered to be well established within the geographic range and not likely to be of conservation concern in the foreseeable future.

At the state level, information is limited. The conservation status of the golden mouse often depends on the location of a state within the geographic range and the extent of field records for the species. NatureServe, a nonprofit conservation organization that manages ecosystem data for the Western Hemisphere, has created a state-by-state conservation ranking for individual species that is used in conjunction with state Natural Heritage (nongame) programs for identifying species of conservation concern (NatureServe 2006). Ranking varies from 1 ("critically impaired") to 5 ("demonstrably widespread, abundant, and secure"). Using this system, the golden mouse is categorized as either ecologically "secure," "apparently secure," or not of specific conservation concern (or "unranked") across much of its range (Patterson et al. 2003; see Figure 7.2). In some states along the range boundary, however, the golden mouse is placed among less secure rankings. Conservation assessment at the individual state level is difficult. Regardless of broader habitat analyses, there exist limited data to predict species presence.

FIGURE 7.2. Conservation status of the golden mouse within each state across its geographic range (dark line) based on NatureServe (2006). Rankings are Secure (S), Apparently Secure (AS), Vulnerable (V), Imperiled (I), Critically Imperiled (CI), and Not Ranked/Under Review (N).

According to state wildlife programs and Gap analysis (Maxwell 2005), golden mouse populations are distributed statewide and considered common in a variety of habitats in Georgia (GDNR 2000, Kramer et al. 2003) and Alabama (ADCNR 2004). The Arkansas Game and Fish Commission has recorded golden mouse presence in every county (Sasse et al. 2001; B. Sasse personal communication). Comparatively, NatureServe designated golden mouse populations in Arkansas as "apparently secure" (NatureServe 2006). Golden mouse populations are considered "secure" in North Carolina (NCWRC 2005) because Gap analysis has identified widespread suitable habitat across the state (NC-GAP 2005). Localized populations of golden mice have been recorded up to at least 830 m in elevation in the Great Smoky Mountains National Park (Linzey 1968b). Interestingly, the North Carolina Wildlife Resources Commission lists the ubiquitous white-footed mouse as a "priority species" for conservation but not the presumably less common golden mouse (NCWRC 2005).

Several states lack consistent information related to the distribution of golden mice. Ranked "secure" in Tennessee by NatureServe (2006), golden mice are considered widespread by the Tennessee Comprehensive Wildlife Conservation Strategy but also noted as "status unknown" because of lack of data or substantially conflicting information (TWRA 2005). Despite widespread documentation of golden mice and their habitat throughout much of Kentucky (Thomas 2001, Wethington et al. 2003) and an "apparently secure" status (NatureServe 2006), the Kentucky Department of Fisheries and Wildlife Resources lists golden mice as uncommon. Linzey and Packard (1977) suggested that the golden mouse range in Virginia included all but the northeastern corner of the state. The Virginia Department of Game and Inland Fisheries (VDGIF 2003), however, suggests that the golden mouse range is limited to the mountainous western portion of the state (Figure 7.3). Although the golden mouse is expected to occur throughout

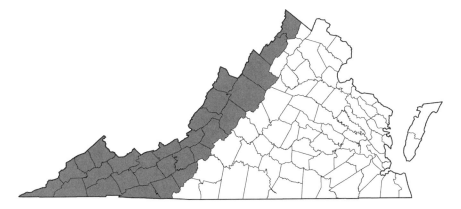

FIGURE 7.3. Range of the golden mouse in Virginia, as estimated by the Department of Game and Inland Fisheries (VDGIF 2003). Lines indicate county boundaries; counties estimated to contain golden mice are shaded. Note the differences between this estimated range map and that described by Packard ([1969] Figure 2.1 of this volume).

Louisiana—where populations are ranked by NatureServe (2006) as "apparently secure"—no current information on its distribution in Louisiana is available.

Despite inconsistencies within individual states, the greatest variability in conservation status of the golden mouse is along the periphery of the geographic range. In Texas, golden mice are ranked as "apparently secure" by NatureServe (2006), although field records indicate that populations are restricted to forests within the eastern part of the state (Davis and Schmidly 1997). Populations are also restricted to the eastern portion of Oklahoma, where the species is ranked as "critically imperiled" by NatureServe (2006). Likewise, range is limited in Missouri, southern Illinois, and southern West Virginia (see Figures 2.1 and 7.2). Conservation status designated by NatureServe (2006) ranges from "vulnerable" in Missouri to "imperiled" in Illinois and West Virginia. In fact, golden mice are designated as "state threatened" in Illinois (Herkert 1992). Although restricted geographically, golden mice have been documented in 22 counties throughout southeastern Missouri (MDC 2006a). However, the Missouri Department of Conservation notes uncertainty with the overall ranking (MDC 2006b). Interestingly, golden mice were identified as state-threatened by the Missouri National Guard (DENIX 1999). Although uncommon or restricted in range within the state, neither the Oklahoma Department of Wildlife Conservation (ODWC 2005) nor the West Virginia Division of Natural Resources (WVDNR 2004) list the golden mouse as "endangered," "threatened," or a "species of concern."

Mississippi, Florida, and South Carolina have a "no ranking" status for golden mice (NatureServe 2006). This might mean that their distribution has not been assessed or that there is conflicting information about their status. Plans for state wildlife management areas in Florida included golden mice as either occurring or expected to occur within six of eight locations: Little Gator Creek, Apalachicola River, Lake Wales Ridge, Chinsegut, Caravelle Ranch, and Joe Budd (FFWCC 2001a, 2002a, 2002b, 2002c, 2002d, 2002e). For these six management areas, no information was provided as to whether golden mice were actually observed on-site. Presence of golden mice was documented at Andrews and Chassahowitzka Wildlife Management Areas (FFWCC 2000, 2001b). Additionally, the presence of golden mouse habitat is common in South Carolina (SCDNR 2006) and the species is common in Mississippi (Shropshire 1998), but information is limited.

Obvious inconsistencies occur in golden mouse conservation status at the state level. For example, the species is listed as "apparently secure" in Texas but as "vulnerable" or "imperiled" within other states along the periphery of the range. Based on principles of biogeography, biophysical factors inherently limit species distribution (Brown and Lomolino 1998), and terrestrial species range boundaries are rarely well aligned with geopolitical boundaries. Thus, it is debatable whether management efforts on the periphery of a species range will actually overcome biophysical variables and enhance overall conservation of golden mice. Compared to well-studied species, such as the white-footed mouse, relatively few data are available about golden mice at the ecosystem or landscape level, although, by means such as Gap analysis, sufficient habitat has been identified as prevalent within the southeastern United States.

Additionally, a drawback of broad habitat analyses such as GAP is that they are "snapshot" analyses of land cover, a disadvantage for species whose habitat might change significantly year to year. In southern Illinois, for example, golden mice have been observed among "atypical" habitats such as autumn olive (*Elaeagnus umbellata*) with little understory and grassy old fields with little woody vegetation (Morzillo 2001). It is also difficult to assess long-term viability of golden mouse populations when associated vegetative communities are replaced over short durations via natural succession or if population presence at a particular location is seemingly ephemeral. For example, Wagner et al. (2000) assessed microhabitat variables of arboreal nests at a site in southern Illinois; there was no evidence of nests there the following year. In another southern Illinois location, arboreal nests used by golden mice (Morzillo et al. 2003) were absent 7 months later.

A disconnect sometimes exists between individual state-level conservation assessments of the golden mouse and those made by broader-level analyses such as NatureServe or Gap analysis. This results in management challenges for conservation practitioners that likely translate to other rare species as well. It is difficult to truly assess the status of the golden mouse. Data on occurrence and abundance are limited, and state-level status assessments can be highly variable. As with other rare species, a broad-spectrum management formula for the golden mouse is difficult to prescribe, and site-specific management is necessary. Natural succession will play a large role in habitat availability, particularly along the range periphery. For active habitat management, activities that retard succession, such as prescribed burning and logging, will help maintain early- to mid-succession vegetation—benefiting golden mice. In southern Illinois, golden mice likely are more abundant now than in the past as a result of previous and current forest management practices that opened the canopy and retarded succession. Regardless of management effort, low local abundance, elusiveness, spatial clustering, and seasonal variation in numbers of this species will likely continue to impact the available data on local populations, with continued challenges to resource managers and conservationists.

Literature Cited

ADCNR (Alabama Department of Conservation and Natural Resources). 2004. Watchable wildlife. Available from http://www.conservation.alabama.gov.

Aldrich, J.W. 1963. Geographic orientation of American Tetraonidae. Journal of Wildlife Management 27:529–545.

Baillie, J. 1996. *Ochrotomys nuttalli*. *In* IUCN 2006. *2006 IUCN Red List of Threatened Species*. Available from www.iucnredlist.org.

Barton, N.H. 1992. The genetic consequences of dispersal. Pages 37–59 *in* N.C. Stenseth and W.Z. Lidicker, editors. Animal dispersal: Small mammals as a model. Chapman & Hall, New York, New York.

Blus, L.J. 1966a. Some aspects of golden mouse ecology in southern Illinois. Transactions Illinois State Academy of Science 59:334–341.

Blus, L.J. 1966b. Relationships between litter size and latitude in the golden mouse. Journal of Mammalogy 47:546–547.

Brown, J.H., and M.V. Lomolino. 1998. Biogeography, 2nd ed. Sinauer Associates, Sunderland, Massachusetts.

Christopher, C.C., and G.W. Barrett. 2006. Coexistence of white-footed mice (*Peromyscus leucopus*) and golden mice (*Ochrotomys nuttalli*) in a southeastern forest. Journal of Mammalogy 87:102–107.

Christopher, C.C., and G.W. Barrett. 2007. Double captures of *Peromyscus leucopus* and *Ochrotomys nuttalli*. Southeastern Naturalist. In press.

Corgiat, D.A. 1996. Golden mouse microhabitat preference: Is there interspecific competition with white-footed mice in southern Illinois? MS thesis, Southern Illinois University, Carbondale, Illinois.

Congreave, P. 1993. The relationship between body size and population abundance in animals. Trends in Ecology and Evolution 8:244–248.

Currie, D.J. 1993. What shape is the relationship between body size and population density? Oikos 66:353–358.

Damuth, J. 1981. Population density and body size in mammals. Nature 290:699–700.

Damuth, J. 1993. Cope's rule, the island rule and the scaling of mammalian population density. Nature 365:748–750.

Davis, W.B., and D.J. Schmidly. 1997. The mammals of Texas. Texas Tech University Press, Lubbock, Texas. Available from http://www.nsrl.ttu.edu/tmot1/.

DENIX (Defense and Environmental Network Information Exchange). 1999. FY1998 Secretary of Defense Environmental Security Awards for Natural Resource Conservation. Available from https://www.denix.osd.mil/denix/Public/News/Earthday99/Awards99/ARMissouri/Missouri_Final.html.

Dueser, R.D. and J.G. Hallett. 1980. Competition and habitat selection in a forest-floor small mammal fauna. Oikos 35:293–297.

Dueser, R.D., and H.H. Shugart. 1978. Microhabitats in a forest-floor small mammal fauna. Ecology 59:89–98.

Durish, N.D., K.E. Halcomb, C.W. Kilpatrick, and R.D. Bradley. 2004. Molecular systematics of the *Peromyscus truei* species group. Journal of Mammalogy 85:1160–1169.

Ebensperger, L.A. 2001. A review of the evolutionary causes of rodent group-living. Acta Theriologica 46:115–144.

Elton, C. 1927. Animal ecology. Sidgwick and Jackson, London, United Kingdom.

Feldhamer, G.A., and K.A. Maycroft. 1992. Unequal capture response of sympatric golden mice and white-footed mice. American Midland Naturalist 128:407–410.

Feldhamer, G.A., and C.R. Paine. 1987. Distribution and relative abundance of the golden mouse (*Ochrotomys nuttalli*) in Illinois. Transactions of the Illinois State Academy of Science 80:213–220.

FFWCC (Florida Fish and Wildlife Conservation Commission). 2000. Conceptual management plan for the Andrews Wildlife and Environmental Area. FFWCC, Tallahassee, Florida.

FFWCC (Florida Fish and Wildlife Conservation Commission). 2001a. Conceptual management plan for the Little Gator Creek Wildlife and Environmental Area. FFWCC, Tallahassee, Florida.

FFWCC (Florida Fish and Wildlife Conservation Commission). 2001b. Conceptual management plan for the Chassahowitzka Wildlife and Environmental Area. FFWCC, Tallahassee, Florida.

FFWCC (Florida Fish and Wildlife Conservation Commission). 2002a. Conceptual management plan for the Apalachicola River Wildlife and Environmental Area. FFWCC, Tallahassee, Florida.

FFWCC (Florida Fish and Wildlife Conservation Commission). 2002b. Conceptual management plan for the Lake Wales Ridge Wildlife and Environmental Area. FFWCC, Tallahassee, Florida.

FFWCC (Florida Fish and Wildlife Conservation Commission). 2002c. Conceptual management plan for the Chinsegut Wildlife and Environmental Area. FFWCC, Tallahassee, Florida.

FFWCC (Florida Fish and Wildlife Conservation Commission). 2002d. Conceptual management plan for the Caravelle Ranch Wildlife and Environmental Area. FFWCC, Tallahassee, Florida.

FFWCC (Florida Fish and Wildlife Conservation Commission). 2002e. Conceptual management plan for the Joe Budd Wildlife and Environmental Area. FFWCC, Tallahassee, Florida.

Gaston, K.J. 1994a. Rarity. Chapman & Hall, New York, New York.

Gaston, K.J. 1994b. Measuring geographic range sizes. Ecography 17:198–205.

Gaston, K.J. 1997. What is rarity? Pages 30–47 in W.E. Kunin and K.J. Gaston, editors. The biology of rarity. Chapman & Hall, London, United Kingdom.

Gaston, K.J., and T.M. Blackburn. 1996. Conservation implications of geographic range size–body size relationships. Conservation Biology 10:638–646.

Gaston, K.J. and W.E. Kunin. 1997. Rare–common differences: An overview. Pages 12–29 in W.E. Kunin and K.J. Gaston, editors. The biology of rarity. Chapman & Hall, London, United Kingdom.

Glazier, D.S. 1980. Ecological shifts and the evolution of geographically restricted species of North American Peromyscus (mice). Journal of Biogeography 7:63–83.

GDNR (Georgia Department of Natural Resources). 2000. Georgia Wildlife. Available from http://museum.nhm.uga.edu/.

Greenwood, P.J. 1980. Mating systems, philopatry and dispersal in birds and mammals. Animal Behaviour 28:1140–1162.

Hall, E.R. 1981. The mammals of North America, 2nd ed. John Wiley & Sons, New York, New York.

Hall, E.R., and K.R. Kelson. 1959. Page 657 in The mammals of North America. Ronald Press, New York, New York.

Herkert, J.R., editor. 1992. Endangered and threatened species of Illinois: Status and distribution. Animals, Vol. 2. Illinois Endangered Species Protection Board, Springfield, Illinois.

Howard, W.E. 1960. Innate and environmental dispersal of individual vertebrates. American Midland Naturalist 63:152–161.

Howell, J.C. 1954. Populations and home ranges of small mammals on an overgrown field. Journal of Mammalogy 35:177–186.

Kaufman, D.W., and G.A. Kaufman. 1989. Population biology. Pages 233–270 *in* G.L. Kirkland, Jr. and J.N. Layne, editors. Advances in the study of *Peromyscus* (Rodentia). Texas Tech University Press, Lubbock, Texas.

Kitchings, T.J., and D.J. Levy. 1981. Habitat patterns in a small mammal community. Journal of Mammalogy 62:814–820.

Knuth, B.A., and G.W. Barrett. 1984. A comparative study of resource partitioning between *Ochrotomys nuttalli* and *Peromyscus leucopus*. Journal of Mammalogy 65:576–583.

Komarek, E.V. 1939. A progress report on southeastern mammal studies. Journal of Mammalogy 20:292–299.

Kramer, E., M.J. Conroy, M.J. Elliott, E.A. Anderson, W.R. Bumback, and J. Epstein. 2003. The Georgia Gap analysis project, Final report. Institute of Ecology and Georgia Cooperative Fish and Wildlife Research Unit, University of Georgia, Athens, Georgia.

Krohne, D.T., B.A. Dubbs, and R. Baccus. 1984. An analysis of dispersal in an unmanipulated population of *Peromyscus leucopus*. American Midland Naturalist 112:146–156.

Kunin, W.E., and K.J. Gaston. 1993. The biology of rarity: Patterns, causes and consequences. Trends in Ecology and Evolution 8:298–301.

Laurance, W.F. 1991. Ecological correlates of extinction proneness in Australian tropical rain forest mammals. Conservation Biology 5:79–89.

Layne, J.N. 1960. The growth and development of young golden mice, *Ochrotomys nuttalli*. Quarterly Journal of the Florida Academy of Science 23:36–58.

Lawton, J.H. 1989. What is the relationship between population density and body size in animals? Oikos 55:429–434.

Lidicker, W.Z., Jr. 1975. The role of dispersal in the demography of small mammals. Pages 103–128 *in* F.B. Golley, K. Petrusewicz, and L. Ryszkowski, editors. Small mammals: Their productivity and population dynamics. Cambridge University Press, New York, New York.

Lidicker, W.Z., Jr., and J.L. Patton. 1987. Patterns of dispersal and genetic structure in populations of small rodents. Pages 144–161 *in* B.D. Chepko-Sade and Z.T. Halpin, editors. Mammalian dispersal patterns. University of Chicago Press, Chicago, Illinois.

Lidicker, W.Z., Jr., and N.C. Stenseth. 1992. To disperse or not to disperse: Who does it and why? Pages 21–36 *in* N.C. Stenseth and W.Z. Lidicker, Jr., editors. Animal dispersal: Small mammals as a model. Chapman & Hall, New York, New York.

Linzey, D.W. 1968a. Mammals of the Great Smoky Mountains National Park. Journal of the Elisha Mitchell Science Society 84:384–414.

Linzey, D.W. 1968b. An ecological study of the golden mouse, *Ochrotomys nuttalli*, in the Great Smoky Mountains National Park. American Midland Naturalist 79:320–345.

Linzey, D.W., and A.V. Linzey. 1967. Growth and development of the golden mouse, *Ochrotomys nuttalli nuttalli*. Journal of Mammalogy 48:445–458.

Linzey, D.W., and R.L. Packard. 1977. *Ochrotomys nuttalli*. Mammalian Species 75:1–6.

Lord, R.D., Jr. 1960. Litter size and latitude in North American mammals. American Midland Naturalist 64:488–499.

Maxwell, J. 2005. The integration of GAP data into state comprehensive wildlife conservation strategies. Gap Analysis Bulletin 13:10–13.

McCarley, W.H. 1958. Ecology, behavior and population dynamics of *Peromyscus nuttalli* in eastern Texas. Texas Journal of Science 10:147–171.

McCarley, W.H. 1959. The effect of flooding on a marked population of *Peromyscus*. Journal of Mammalogy 40:57–63.

McCoy, E.D., and E.F. Connor. 1980. Latitudinal gradients in the species density of North American mammals. Evolution 34:193–203.

McDonald, L.L. 2004. Sampling rare populations. Pages 11–42 *in* W.L. Thompson, editor. Sampling rare or elusive species. Island Press, Washington, DC.

MDC (Missouri Department of Conservation). 2006a. Missouri Fish and Wildlife Information System. MDC, Columbia, Missouri. Available from http://mdc4.mdc.mo.gov/applications/mofwis/mofwis_search1.aspx.

MDC (Missouri Department of Conservation). 2006b. Missouri species and communities of conservation concern: Checklist January 2006. MDC, Columbia, Missouri.

Miller, D.A., R.E. Thill, M.A. Melchiors, T.B. Wigley, and P.A. Tappe. 2004. Small mammal communities of streamside management zones in intensively managed pine forests of Arkansas. Forest Ecology and Management 203:381–393.

Morzillo, A.T. 2001. Nest and habitat use by the golden mouse (*Ochrotomys nuttalli*) in southern Illinois. MS thesis, Southern Illinois University, Carbondale, Illinois.

Morzillo, A.T., G.A. Feldhamer, and M.C. Nicholson. 2003. Home range and nest use of the golden mouse (*Ochrotomys nuttalli*) in southern Illinois. Journal of Mammalogy 84:553–560.

Munton, P. 1987. Concepts of threat to the survival of species used in the Red Data books and similar compilations. Pages 72–95 *in* R. Fitter and M. Fitter, editors. The road to extinction. IUCN/UNEP, Gland, Switzerland.

NatureServe. 2006. NatureServe Explorer: An online encyclopedia of life [web application]. Version 4.7. NatureServe, Arlington, Virginia. Available from http://www.natureserve.org/explorer.

Nadeau, J.H., R.T. Lombardi, and R.H. Tamarin. 1981. Population structure and dispersal of *Peromyscus leucopus* on Muskeget Island. Canadian Journal of Zoology 59:793–799.

NC-GAP Analysis Project. 2005. North Carolina species report: Golden mouse. Department of Zoology, North Carolina State University, Raleigh, North Carolina.

NCWRC (North Carolina Wildlife Resources Commission). 2005. North Carolina Wildlife Action Plan. Raleigh, North Carolina.

ODWC (Oklahoma Department of Wildlife Conservation). 2005. Oklahoma Comprehensive Wildlife Conservation Strategy. ODWC, Oklahoma City, Oklahoma.

Packard, R.L. 1969. Taxonomic review of the golden mouse, *Ochrotomys nuttalli*. Miscellaneous Publication 51, Pages 373–406. University of Kansas Museum of Natural History, Lawrence, Kansas.

Pagel, M.D., R.M. May, and A.R. Collie. 1991. Ecological aspects of the geographic distribution and diversity of mammalian species. American Naturalist 137:791–815.

Patterson, B.D., G. Ceballos, W. Sechrest, M.F. Tognelli, T. Brooks, L. Luna, P. Ortega, I. Salazar, and B.E. Young. 2003. Digital distribution maps of the mammals of the Western Hemisphere, Version 1.0. NatureServe, Arlington, Virginia.

Pearson, P.G. 1953. A field study of *Peromyscus* populations in Gulf Hammock, Florida. Ecology 34:199–207.

Pimm, S.L., and C. Jenkins. 2005. Sustaining the variety of life. Scientific American 293:66–73.

Pruett, A.L., C.C. Christopher, and G.W. Barrett. 2002. Effects of a forested riparian peninsula on mean home range size of the golden mouse (*Ochrotomys nuttalli*) and the white-footed mouse (*Peromyscus leucopus*). Georgia Journal of Science 60:201–208.

Rabinowitz, D. 1981. Seven forms of rarity. Pages 205–217 *in* H. Synge, editor. The biological aspects of rare plant conservation. John Wiley & Sons, Chichester, United Kingdom.

Rabinowitz, D., S. Cairns, and T. Dillon. 1986. Seven forms of rarity and their frequency in the flora of the British Isles. Pages 182–204 *in* M.E. Soule, editor. Conservation biology, the science of scarcity and diversity. Sinauer Associates, Sunderland, Massachusetts.

Rapoport, E.H. 1982. Areography: Geographical strategies of species. Pergamon Press, Oxford, United Kingdom.

Sasse, B., K. Rowe, N. Childers, G. Heidt, T. Nupp, and D. Saugey. 2001. Strategic nongame mammal management plan. Arkansas Game and Fish Commission, Little Rock, Arkansas.

SCDNR (South Carolina Department of Natural Resources). 2006. South Carolina Gap analysis project. Available from http://www.dnr.sc.gov/GIS/gap/scgaphome.html.

Schoener, T.W. 1987. The geographical distribution of rarity. Oecologia 74:161–173.

Seagle, S.W. 1985. Competition and coexistence of small mammals in an east Tennessee pine plantation. American Midland Naturalist 114:272–282.

Shadowen, H.E. 1963. A live-trap study of small mammals in Louisiana. Journal of Mammalogy 44:103–108.

Shropshire, C.C. 1998. Land mammals of Mississippi. Mississippi Department of Wildlife, Fisheries, and Parks, Jackson, Mississippi.

Solomon, N.G. 2003. A reexamination of factors influencing philopatry in rodents. Journal of Mammalogy 84:1182–1197.

Stenseth, N.C., and W.Z. Lidicker, Jr., editors. 1992. Animal dispersal: Small mammals as a model. Chapman & Hall, New York, New York.

Stevens, G.C. 1989. The latitudinal gradient in geographical range: How so many species coexist in the tropics. American Naturalist 133:240–256.

Thomas, S.C. 2001. The statewide small mammal survey, Final report. Kentucky Department of Fish and Wildlife Resources, Frankfort, Kentucky.

Thompson, W.L., editor. 2004. Sampling rare or elusive species: Concepts, designs, and techniques for estimating population parameters. Island Press, Washington, DC.

TWRA (Tennessee Wildlife Resources Agency). 2005. Tennessee's Comprehensive Wildlife Conservation Strategy. TWRA, Nashville, Tennessee.

VDGIF (Virginia Department of Game and Inland Fisheries). 2003. Virginia wildlife information: Common golden mouse (*Ochrotomys nuttalli aureolus*). Richmond, Virginia. Available from http://www.dgif.state.va.us/wildlife/.

Wagner, D.M., G.A. Feldhamer, and J.A. Newman. 2000. Microhabitat selection by golden mice (*Ochrotomys nuttalli*) at arboreal nest sites. American Midland Naturalist 144:220–225.

Wethington, K., T. Derting, T. Kind, H. Whiteman, M. Cole, M. Drew, D. Fredricks, G. Ghitter, A. Smith, and M. Soto. 2003. The Kentucky GAP analysis project, Final report. Kentucky Department of Fish and Wildlife Resources, Frankfort, Kentucky.

Wolff, J.O. 1989. Social behavior. Pages 271–291 *in* G.L. Kirkland, Jr. and J.N. Layne, editors. Advances in the study of *Peromyscus* (Rodentia). Texas Tech University Press, Lubbock, Texas.

Wolff, J.O., K.I. Lundy, and R. Baccus. 1988. Dispersal, inbreeding avoidance, and reproductive success of white-footed mice. Animal Behaviour 36:456–465.

WVDNR (West Virginia Wildlife Diversity Program). 2004. Mammals of West Virginia Field Checklist. Wildlife Diversity Program, Elkins, West Virginia.

Yu, J., and F.S. Dobson. 2000. Seven forms of rarity in mammals. Journal of Biogeography 27:131–139.

8

The Golden Mouse: A Model of Energetic Efficiency

JOHN D. PELES AND GARY W. BARRETT

Those systems that survive in the competition among alternative choices are those that develop more power inflow and use it best to meet the needs of survival (H.T. Odum and E.C. Odum 1976:40)

Introduction

Patterns of energy acquisition and use (i.e., bioenergetics) by mammals have important consequences for the ecology of the individual organism as well as the populations, communities, ecosystems, and landscapes in which they live (Ernest 2005, Grodzinski and Wunder 1975, McNab 1980, 2002). A bioenergetic pathway for a typical mammal is depicted in Figure 8.1 and can be summarized in the form of three equations:

(1) $I = E + A$, where I is ingested energy, E is energy lost by the egestion of nondigestable material in the form of feces, and A is assimilated energy that is available to the organism following digestion and absorption;

(2) $A = U + M + G + R$ represents the energy budget of the organism (McNab 2002), where U is energy lost through the excretion of nitrogenous waste products in urine and M, G, and R represent energy allocated for metabolism, growth, and reproduction, respectively;

(3) $M = h + m + t + a$, where h is energy lost to the organism in the form of heat during cellular respiration, m is energy needed for the maintenance of basal metabolic rate, t is energy used for thermoregulation, and a is energy used for activities such as foraging, predator avoidance, and dispersal.

Values for bioenergetic parameters are typically expressed in the form of a rate such as the quantity of energy per unit body mass per unit time (e.g., kcal · g live weight^{-1} · day^{-1}).

Although all mammalian species use energy for the same basic processes (Figure 8.1), individual species differ with respect to the ingestion and use of energy as a function of life history, behavior, and physiology (McNab 2002,

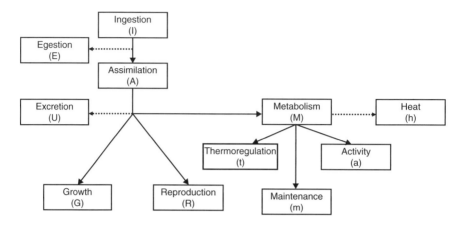

FIGURE 8.1. Pathway of energy flow through a typical mammalian species. Dashed lines indicate processes involving the loss of usable energy to the organism. The relative proportion of assimilated energy that is allocated for each function will depend on environmental conditions and differences in life history, behavior, and physiology of the species.

Tomasi and Horton 1992). For example, ingestion of energy (I) will vary due to differences in competitive abilities, foraging behaviors, and dietary preferences among species. Likewise, physiological differences among species will result in differences in energy use that can be assessed using two bioenergetic parameters:

(1) Assimilation efficiency ($\%A$) = $A \cdot I^{-1}$ where $\%A$ is a measure of the proportion of ingested energy that is assimilated through the processes of digestion and absorption;

(2) Net production efficiency ($\%P$) = $(G + R) \cdot A^{-1}$, where $\%P$ is a measure of the proportion of assimilated energy that is available for growth and reproduction following allocation of energy for metabolism (M).

The survival and reproduction of an organism are dependent on adaptations that permit the maximization of energy consumption and/or the efficient use of energy to ensure its availability for critical processes, especially during periods of energetic stress (McNab 2002, Tomasi and Horton 1992). For small mammals that inhabit seasonally cold environments, significant increases in energetic demands for thermoregulation occur during winter (Feldhamer et al. 2004). In these environments, increased $\%A$ and $\%P$ would be favorable prior to the onset of cold temperatures to maximize the production of brown fat used in nonshivering thermogenesis (Feldhamer et al. 2004). Likewise, increased $\%A$ during cold periods would increase the amount of energy available for metabolic heat production. Energetic efficiency in small mammals is also important during reproductive periods when increased amounts of energy must be allocated by females for production of offspring (Gittleman and Thompson 1988, Thompson 1992).

Although all small mammals possess adaptations for acquiring and using energetic resources, we suggest that the golden mouse (*Ochrotomys nuttalli*) represents a model of energetic efficiency. Increased energetic efficiency of this species not only permits golden mice to meet the energetic demands of survival and reproduction in a temperate environment, but also has important consequences for ecological relationships with other species such as the white-footed mouse (*Peromyscus leucopus*). *O. nuttalli* and *P. leucopus* coexist throughout the entire range of the golden mouse and share many characteristics, including body size, habitat preferences, food preferences, and periods of activity (Christopher and Barrett 2006, Knuth and Barrett 1984, Lackey 1985, Linzey and Packard 1977). In this chapter, we will (1) summarize the physiological and bioenergetic characteristics of golden mice, (2) discuss the relationship of behavioral characteristics of golden mice to bioenergetics, and (3) present and discuss an overall model of the bioenergetic strategy of golden mice that illustrates the energetic efficiency of this species and its significance to its ecology. The bioenergetic efficiency of golden mice will also be compared with *P. leucopus* and discussed in the context of coexistence between these two species.

Physiology and Bioenergetics

Because maintenance of body temperature and basal metabolic rate represents potentially significant uses of energy for endothermic species (Figure 8.1), knowledge concerning these physiological characteristics is important for understanding the bioenergetics of golden mice. A comprehensive assessment of physiological characteristics related to thermoregulation and metabolism in golden mice was conducted by Layne and Dolan (1975). In that investigation, mice maintained a mean body temperature of 36.38°C and a basal metabolic rate of 1.30 mL O_2/g/h between 5°C and 35°C. Based on comparisons with reported values for several species of *Peromyscus*, including *P. leucopus*, it was concluded that *O. nuttalli* has comparable thermoregulatory abilities to *Peromyscus* but tends to have a lower body temperature and metabolic rate (Layne and Dolan 1975). Further evidence for physiological differences between *O. nuttalli* and *P. leucopus* was provided by a comparison of body temperatures of both species at two different ambient temperatures. For mice maintained at 10°C and 20°C, the mean body temperatures of *O. nuttalli* (36.55°C and 36.33°C) were significantly lower compared to *P. leucopus* (37.93°C and 37.17°C [Knuth and Barrett 1984]).

Several investigators have reported bioenergetic values for golden mice under a variety of conditions within a laboratory setting (Table 8.1). Results of these investigations demonstrate that differences in dietary quality might influence either the amount of ingested energy (*I*) or assimilation efficiency (%A). For example, ingestion values for mice fed smooth sumac (*Rhus glabra*)—a food source high in phenolic compounds—were an order of magnitude lower than for

TABLE 8.1. Values for ingestion (*I*), egestion (*E*), assimilation (*A*), and assimilation efficiency (%*A*) of golden mice under differing dietary conditions.

Diet	*I*	*E*	*A*	%*A*	Source
75% Lab chow/					
25% sunflower seeds	0.80	0.16	0.64	80.0	Stueck et al. (1977)
50% Lab chow/					
50% sunflower seeds	0.75	0.10	0.57	86.7	Stueck et al. (1977)
25% Lab chow/					
75% sunflower seeds	0.62	0.05	0.57	91.9	Stueck et al. (1977)
Sunflower seeds	0.50	0.02	0.48	96.0	Springer et al. (1981)
Sunflower seeds	0.41*	0.02*	0.39*	95.1*	Springer et al. (1981)
Sunflower seeds	0.86**	0.03**	0.83**	96.6**	Knuth and Barrett (1984)
Sunflower seeds	0.61	0.02	0.59	96.7	Knuth and Barrett (1984)
Smooth sumac					
(recently matured)	0.03	0.01	0.03	86.7	Jewell et al. (1991)
Smooth sumac					
(1-year old)	0.09	0.02	0.07	76.6	Jewell et al. (1991)
Japanese honeysuckle					
berries	0.95	0.08	0.87	92.0	Peles et al. (1995)
Eastern red cedar					
berries	0.42	0.01	0.41	98.0	Peles, et al. (1995)
Water oak acorns/					
privet berries	1.06	0.23	0.83	78.3	O'Malley et al. (2003)

Values for *I*, *E*, and *A* are reported as kcal · g live weight^{-1} · day^{-1}. Unless otherwise indicated, all mean values were determined for mice maintained individually (ungrouped) at 20–22°C.
*Values determined for mice maintained as groups of three individuals at 22°C.
**Values determined for mice maintained individually at 10°C.

those reported in any other study (Jewell et al. 1991). In a feeding study involving two potential food sources commonly found in the natural habitat of golden mice, Peles et al. (1995) reported values of ingestion and assimilation for mice fed fruits and seeds of Japanese honeysuckle (*Lonicera japonica*) that were twice as high as those on a diet of Eastern red cedar (*Juniperus virginiana*). Because these food sources differed little with respect to protein and caloric content, or with respect to the presence of phenolic compounds, consumption was likely influenced by some aspect of palatability (Peles et al. 1995).

In most cases, the amount of ingested energy is highly correlated with assimilated energy (Table 8.1) and this determines the amount of energy available for allocation to metabolism, growth, and reproduction. However, assimilation efficiencies in small mammals might vary from 65 percent to 95 percent as a function of dietary quality (Drozdz 1968). Most investigators have reported relatively high values (> 90 percent) for assimilation efficiency of golden mice fed both naturally occurring food sources and laboratory diets (Table 8.1). Lower (< 90 percent) assimilation efficiencies reported by Jewell et al. (1991) and O'Malley et al. (2003) likely reflect the presence of high levels of dietary phenolic compounds such as tannins. Interestingly, Stueck et al. (1977) observed that assimilation efficiency decreased as a function of increasing caloric content of the diet.

Golden mice also alter rates of ingestion to compensate for changes in caloric needs imposed by environmental conditions (Table 8.1). For example, Knuth and Barrett (1984) found that values of ingestion and assimilation were both significantly greater for golden mice maintained at 10°C compared to individuals fed the same diet at 20°C. However, percent assimilation efficiency did not differ between temperatures. Springer et al. (1981) reported significantly greater ingestion and assimilation values in golden mice housed individually compared to those housed in groupings of three mice. They concluded that these differences reflect the benefits of communal nesting on the costs of thermoregulation.

Two of the investigations summarized in Table 8.1 compared bioenergetic values between *O. nuttalli* and *P. leucopus*. Knuth and Barrett (1984) found that golden mice had significantly lower rates of ingestion and assimilation compared to *P. leucopus* at both 10°C and 20°C. Although assimilation efficiencies were relatively high (> 95 percent) in both species at each temperature, golden mice (96.70 percent) exhibited a significantly greater assimilation efficiency compared to *P. leucopus* (95.39 percent) at 20°C. O'Malley et al. (2003) also reported significantly lower rates of ingestion and assimilation in golden mice compared to *P. leucopus*. Unlike Knuth and Barrett (1984), they observed significantly lower assimilation efficiency for golden mice (73.50 percent) compared to *P. leucopus* (90.40 percent). However, these findings were confounded by differences in dietary quality between the two species. For example, *P. leucopus* selected from a diet of water oak (*Quercus nigra*) acorns, white oak (*Q. alba*) acorns, and Chinese privet (*Ligustrum sinense*) berries, whereas *O. nuttalli* were fed a diet of only water oak acorns and privet berries.

Golden mice are generally more energetically efficient than *P. leucopus* (Knuth and Barrett 1984, O'Malley et al. 2003). Assuming no differences in dietary quality (i.e., similar caloric content, palatability, and biochemical composition), *P. leucopus* must consume and assimilate more energy to meet the energetic requirements associated with maintenance of a higher basal metabolic rate and body temperature (Figure 8.2). In contrast, maintenance of a comparatively lower body temperature and metabolic rate in *O. nuttalli* (Knuth and Barrett 1984, Layne and Dolan 1975) requires less ingested energy. The hypothetical energy flow diagrams depicted in Figure 8.2 are based on the assumption of similar assimilation efficiencies between species. However, based on the findings of Knuth and Barrett (1984), a higher percentage of assimilated energy might be available for metabolism, growth, and maintenance in golden mice compared to *P. leucopus*. This increased assimilation efficiency, coupled with reduced energetic requirements of golden mice, would provide a potentially important competitive advantage for this species in environments that might be energetically limited (McNaughton and Wolf 1970) or during energetically stressful times such as reproduction (e.g., Thompson 1992).

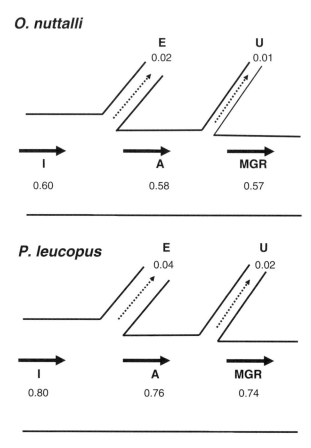

FIGURE 8.2. Hypothetical energy flow diagrams for a golden mouse (*O. nuttalli*) and white-footed mouse (*P. leucopus*) fed a high-quality diet at 20°C. Bioenergetic (*I* = ingested energy; *E* = energy lost by egestion of nondigestible material in the form of feces; *A* = assimilated energy; *U* = energy lost through the excretion of nitrogenous waste products in urine; MGR = energy available for metabolism, growth, and reproduction) values are based on data from Stueck et al. (1977) and Knuth and Barrett (1984). Bioenergetic values are expressed as kcal · g live weight^{-1} · day^{-1}.

Behavior and Bioenergetics

Most studies concerned with the energetics of small mammals have involved the examination of physiological characteristics (body temperature, basal metabolic rate) or the determination of bioenergetic values (*I, A, %A*). Little attention has been given to the association of these characteristics with behavior. Perhaps the best example of the importance of considering the relationship of behavior with bioenergetics is provided by the study of communal nesting behavior in golden mice.

Although many species of small mammals exhibit communal nesting as a means of reducing the energetic costs of thermoregulation (Feldhamer et al. 2004), the use of this behavior is exemplified by golden mice (Dietz and Barrett 1992, Springer et al. 1981). Linzey and Packard (1977) described the golden mouse as a "fairly social" animal and several investigators have reported as many as eight adult golden mice in the same nest under natural conditions (Goodpaster and Hoffmeister 1954, Ivey 1949, Linzey and Packard 1977, Springer et al. 1981). One of us (G. W. Barrett) has observed six to eight golden mice in a single nest on multiple occasions in a population of mice near Bighill, Madison County, Kentucky. Furthermore, Dietz and Barrett (1992) found no difference in the frequency of kin versus nonkin in communal nests under experimental conditions. Thus, golden mice might form large social groups of unrelated individuals (Dietz and Barrett 1992).

The importance of communal nesting on bioenergetics of golden mice has been demonstrated by Springer et al. (1981). They found that golden mice nesting in experimental groups of three individuals ingested significantly less energy than mice nesting alone. Mice were provided with unlimited amounts of high-quality food in both treatments and no changes in body size or reproductive status were observed. Consequently, the decreased rates of energy intake by grouped mice were attributable to the effects of communal nesting on the energetic costs of thermoregulation (Springer et al. 1981). In this scenario, communal nesting complements the effects of relatively low body temperature and low basal metabolic rate to increase the energetic efficiency of golden mice.

Numerous investigators have commented on the docile nature and lack of aggressiveness of golden mice compared to other species of small mammals (Christopher and Barrett 2006, Goodpaster and Hoffmeister 1954, Ivey 1949, Linzey and Packard 1977, McCracken 1978). Their docile behavior is likely an important factor that is conducive to communal nesting. In addition, docile behavior and lack of aggressiveness also would represent an energy conservation mechanism that minimizes energetic requirements associated with activity. Support for this notion is provided by the fact that Ivey (1949) reported that captive golden mice do not use an exercise wheel and rarely engage in unnecessary activity.

Selection for either the type or quantity of food represents another behavior that might influence bioenergetics of golden mice. Stueck et al. (1977) found that golden mice fed varying mixtures of laboratory chow and sunflower seeds (Table 8.1) decreased rates of ingestion with increasing caloric content of the diet. Golden mice also alter feeding preferences based on other aspects of the food item that affect quality. In a study of food selection, Knuth and Barrett (1984) found that golden mice select a variety of food sources over staghorn sumac (*Rhus typhina*). Sumac seeds are reported as a potential food source in many habitats occupied by golden mice (Linzey and Packard 1977). Despite the abundance of sumac, mice appear to avoid this food because it contains high levels of dietary phenolic compounds and leads to decreased efficiency of assimilation (Table 8.1; see Jewell et al. 1991). In another study in which golden mice were given the choice of two food items, water oak acorns accounted for 91 percent of the calories in the diet compared to 9 percent for Chinese privet berries (O'Malley et al. 2003). Water oak acorns are nearly six times higher in phenolic content than privet berries, but they also contain twice as many calories. Thus, golden mice appear to have selected for caloric content at the expense of reduced assimilation efficiency (Table 8.1) due to the presence of phenolic compounds (O'Malley et al. 2003).

A Model of Energetic Efficiency

The golden mouse is a model species for the study of energetic efficiency in an endotherm. This energetic efficiency results from numerous interrelationships among bioenergetic, physiological, and behavioral characteristics (Figure 8.3) described in this chapter. Energetic costs for metabolism (*M*) are reduced by the

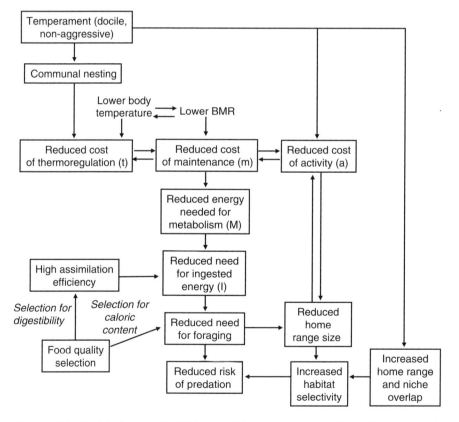

FIGURE 8.3. Model of energetic efficiency and its consequences in the golden mouse (*O. nuttalli*). The interrelationships between bioenergetic and behavioral or physiological characteristics of golden mice are depicted. Possible associations of these characteristics with ecologically relevant factors are also shown. BMR represents basal metabolic rate.

effects of behavioral and physiological characteristics on all three categories of metabolic energy needs (t, a, and m [Figure 8.1]). As described in the previous section, the temperament of the golden mouse (Linzey and Packard 1977) enhances the ability of this species to nest communally. Communal nesting and the ability to maintain a body temperature that is somewhat lower than similar species of small mammals, such as *P. leucopus* (Knuth and Barrett 1984, Layne and Dolan 1975), reduce the cost of thermoregulation (t). Docile temperament, lack of aggressive behavior, and minimization of unnecessary activity also reduce the energetic costs associated with activity (a). Finally, lower basal metabolic rate of golden mice relative to other species of mice (Layne and Dolan 1975) reduces the energetic costs of maintenance (m).

Reduced intake of energy (I) in golden mice compared to *P. leucopus* has been well documented (Knuth and Barrett 1984, O'Malley et al. 2003) and likely reflects the reduced metabolic energy needs of this species (Figure 8.3). Whereas behaviors such as communal nesting reduce the need for ingested energy by lowering energetic requirements for maintenance, the need for ingested energy is further reduced by the high rates of assimilation (%A) that have commonly been observed in golden mice (Knuth and Barrett 1984, Peles et al. 1995, Springer et al. 1981, Stueck et al. 1977 [Table 8.1]). Assimilation efficiency might be further facilitated by food selection behavior for food items that are relatively low in dietary phenolic compounds (Figure 8.3). The effect of food selection on digestibility and assimilation efficiency might be offset in some cases by selection for caloric content (O'Malley et al. 2003). However, as will be discussed in the next section, food selection for caloric content has important implications for the ecology of golden mice.

Ecological Consequences of Energetic Efficiency

The energetic efficiency of golden mice and the factors that contribute to this efficiency likely influence many aspects of the ecology of this species (Figure 8.3). Perhaps the most significant influence of energetic efficiency in golden mice is that reduced need for ingested energy reduces the amount of space (i.e., home range size) needed for foraging (Figure 8.3). Traditionally, the two-dimensional home range of golden mice has been considered relatively small compared to other species of small mammals (Christopher and Barrett 2006, Morzillo et al. 2003). Although three-dimensional home ranges of golden mice are probably larger than those previously determined from a two-dimensional perspective (Christopher and Barrett 2006, Morzillo et al. 2003), the home range of this species is still smaller compared to other small mammal species (Christopher and Barrett 2006). Meserve (1977) suggested that home ranges of small mammals reflect the bioenergetic needs of the species. This is likely the case for golden mice. It should also be noted that there exists a positive feedback relationship (Figure 8.3) between home range and energetic requirements because a reduction in foraging activity and home range use would result in lower energetic costs associated with activity (a).

Although some investigators have suggested that golden mice might be less habitat-specific than previously believed (Knuth and Barrett 1984, Morzillo et al. 2003, Seagle 1985), this species is generally considered to be an extreme habitat specialist (Christopher and Barrett 2006, Dueser and Hallett 1980, Dueser and Shugart 1979, Frank and Layne 1992, Linzey and Packard 1977, Seagle 1985) compared to *P. leucopus*. Energetic efficiency is one factor that could facilitate habitat specificity (Figure 8.3). Reduced home range requirements resulting from increased energetic efficiency and decreased energy requirements permit golden mice to be highly selective with respect to foraging site and nest site location. In addition, space requirements are further reduced by the docile, nonaggressive nature of the species (Goodpaster and Hoffmeister 1954, Ivey 1949, Linzey and Packard 1977, McCracken 1978) that results in an increased overlap of individual home ranges and minimal territoriality (Morzillo et al. 2003). This sharing of space, coupled with the efficient use of food resources, permits maximal use of preferred habitat space at the population level.

In addition to a reduction in the amount of space needed for foraging, reduced energy requirements also minimize the amount of time needed for foraging. Because the amount of time spent foraging is directly related to the risk of predation, the reduced need for ingested energy consequently reduces the risk of predation (Knuth and Barrett 1984). Risk of predation is further minimized by reduced home range requirements (resulting from reduced energy and foraging requirements) that permit golden mice to select microhabitats of heavy undergrowth that provide protection from predators (Linzey and Packard 1977, O'Malley et al. 2003, Wagner et al. 2000). Microhabitats consisting of thick vegetation chosen by golden mice for arboreal nest sites likely provide an advantage for thermoregulation as well (Wagner et al. 2000).

Although seldom considered in studies of small mammal community ecology, differences in use of energy resources might have a significant influence on habitat use and niche partitioning. In most cases, the limiting factor for communities that include *O. nuttalli* and species of *Peromyscus* is probably not food (Christopher and Barrett 2006, Young and Stout 1986) but space for foraging and nest sites (Christopher and Barrett 2006, Frank and Layne 1992). Energetic differences between golden mice and *P. leucopus* permit partitioning of spatial resources and coexistence between the two species. The sociality of golden mice appears to play a positive role in this relationship. The golden mouse is not only considered an extreme habitat specialist, but it is a superior competitor in its preferred habitat, which often might represent a relatively small proportion of the available habitat space (Dueser and Hallett 1980, Dueser and Shugart 1979). Our model of the relationships between energetics and ecology in this species clearly demonstrates how energetic efficiency would provide an advantage in this setting. In contrast, *P. leucopus* is a generalist that uses a wide range of microhabitats (Frank and Layne 1992), which is reflective of its higher metabolic rate (Layne and Dolan 1975) and greater need for ingested energy (Knuth and Barrett 1984, O'Malley et al. 2003).

Energetic efficiency is also likely related to the ability of golden mice to persist in habitats throughout its range. Christopher and Barrett (2006) found that when *P. leucopus* were experimentally removed from a community, golden mice respond only by increasing their use of microhabitats in lower levels of the canopy. This minimal expansion of habitat use suggests that golden mice are not limited to specific microhabitats by competition but by aspects of life history or physiology. Thus, energetic efficiency permits persistence of golden mice in small mammal communities by conferring superior competitive ability in very specific microhabitats (Dueser and Hallett 1980, Dueser and Shugart 1979).

Future Research Challenges

In this chapter, we have examined the causes and consequences of energetic efficiency in the golden mouse. There are a number of interesting species-specific questions that remain to be addressed in future investigations. For example: What is the diet of golden mice in natural populations; and what are the bioenergetic implications of differences in diet and food quality between golden mice and other species in a community of small mammals? There exist a number of additional questions that might be addressed using golden mice as a model species that have broader implications for the study of mammalian energetics and ecology. We will conclude by proposing three questions that we suggest should be the focus of future investigations involving *O. nuttalli* and their relationship to other species of small mammals.

(1) *What is the influence of energetic efficiency on net secondary production efficiency?* Relatively few investigations (Kaczmarski 1966, Liu et al. 2003, Mattingly and McClure 1985) have determined bioenergetic values for small mammals during reproduction. Because most bioenergetic studies are usually limited to less than 1 month, energy allocation for growth and reproduction is usually considered to be zero and net secondary production efficiency (%*P*) is not assessed. In an organism such as the golden mouse, in which the energetic cost of metabolism (*M*) is lower compared to another organism of similar body mass (e.g., *P. leucopus*), there are two possible responses of the organism with lower energetic costs. The first possibility is that it will consume less energy (*I*). As we have discussed in this chapter, this response has been well documented in bioenergetic studies of golden mice (Knuth and Barrett 1984, O'Malley et al. 2003) and there are numerous ecological benefits associated with this response (Figure 8.3).

A second possible response to reduced energy requirements is that the first organism (*O. nuttalli*) consumes the same amount of energy as the second (*P. leucopus*). In this case, the first organism will have comparatively more energy available for growth (G) and reproduction (R); see Figure 8.1. This response would provide a potential selective advantage for golden mice during energetically costly periods of reproduction (Gittleman and Thompson 1988, Thompson 1992). We suggest that future bioenergetic studies of golden mice and other species of small mammals be designed to evaluate changes in food consumption and energy allocation patterns over periods of time that include reproduction.

Comparative studies of net secondary production efficiency might be especially applicable to explaining the coexistence of species such as *O. nuttalli* and *P. leucopus*. Linzey and Packard (1977) concluded that golden mouse litter size tends to be smaller compared to species such as *P. leucopus*. Because golden mice are well adapted to very specific habitats, space is limited and reduced litter size might serve as a regulatory mechanism for population size. Thus, it might be predicted that even though golden mice would increase ingestion rates to compensate for the energetic costs of reproduction, the secondary production efficiency of this species would not differ from *P. leucopus*. This hypothesis deserves further investigation under field conditions.

(2) *What is the association between protein heterozygosity and energetic efficiency*? Numerous investigators have documented associations between multi-locus protein heterozygosity (H) and fitness-related characteristics such as growth, reproduction, secondary sex characteristics, and survivorship in a variety of species, including mammals (reviewed by Mitton 1993; Mitton and Grant 1984). The underlying mechanism for these associations has commonly been attributed to the effects of H on metabolic efficiency and this has been documented in several ectothermic species (Mitton 1993, Mitton and Grant 1984). In this scenario, individuals with increased H have a reduced metabolic rate and, therefore, greater amounts of energy for allocation to growth and reproduction. Only two investigations to date have examined the association between H and metabolic rate in an endothermic species (Carter et al. 1999, Peles and Merritt 2005) and the results of these investigations were opposite the findings for ectothermic species. Carter et al. (1999), for example, found that the metabolic rate of laboratory mice (*Mus domesticus*) subjected to forced exercise was greater in individuals with higher levels of H. Peles and Merritt (2005) documented a similar relationship in northern short-tailed shrews (*Blarina brevicauda*) subjected to thermal stress. Future studies should be designed to elucidate the influences of H on energetics in small mammals and to understand the relevance of these influences to the fitness of the organism. Comparative studies involving golden mice and *P. leucopus* would be especially interesting, given the differing bioenergetic and ecological strategies of these two small mammal species.

(3) *How does bioenergetics influence ecological processes and relationships among mammalian species at the community and ecosystem levels*? Previous investigators have considered the influence of energetic requirements and energy use in small mammals on population ecology (McNab 1980) and community structure (Ernest 2005). In addition, the importance of small

mammal energetics to ecosystem-level processes has been addressed (Grodzinski and Wunder 1975). However, the influence of bioenergetics on ecologically important processes such as predation, competition, and niche partitioning has received little attention. In this chapter, we have proposed a number of ecological processes involving the golden mouse that are likely influenced by its pattern of energy acquisition and use. Specifically, we have proposed that differences in energy use between *O. nuttalli* and *P. leucopus* permit coexistence between these two species. We propose that future investigations test hypotheses concerning bioenergetics based on knowledge of ecological relationships among species. For example, based on knowledge of microhabitat use, home range, and feeding habits, hypotheses concerning bioenergetics could be tested for coexisting species such as *O. nuttalli–P. gossypinus*, *P. leucopus–P. maniculatus*, and *Sigmodon hispidus–Oryzomys palustris*.

Literature Cited

Carter, P.A., T. Garland, M.R. Dohm, and J.P. Hayes. 1999. Genetic variation and correlations between genotype and locomotor physiology in outbred laboratory house mice (*Mus domesticus*). Comparative Biochemistry and Physiology Series A 123:155–162.

Christopher, C.C., and G.W. Barrett. 2006. Coexistence of white-footed mice (*Peromyscus leucopus*) and golden mice (*Ochrotomys nuttalli*) in a southeastern forest. Journal of Mammalogy 87:102–107.

Dietz, B.A., and G.W. Barrett. 1992. Nesting behavior of *Ochrotomys nuttalli* under experimental conditions. Journal of Mammalogy 73:577–581.

Drozdz, A. 1968. Digestibility and assimilation of natural foods in small rodents. Acta Theriologica 13:367–389.

Dueser, R.D., and J.G. Hallett. 1980. Competition and habitat selection in a forest-floor small mammal community. Oikos 35:293–297.

Dueser, R.D., and H.H. Shugart, Jr. 1979. Niche pattern in a forest-floor small-mammal fauna. Ecology 60:108–118.

Ernest, S.K.M. 2005. Body size, energy use, and community structure of small mammals. Ecology 86:1407–1413.

Feldhamer, G.A., L.C. Drickamer, S.H. Vessey, and J.F. Merritt. 2004. Mammalogy: Adaptation, diversity, and ecology, 2nd ed. McGraw-Hill, New York, New York.

Frank, P.A., and J.N. Layne. 1992. Nests and daytime refugia of cotton mice (*Peromyscus gossypinus*) and golden mice (*Ochrotomys nuttalli*) in south-central Florida. American Midland Naturalist 127:21–30.

Gittleman, J.L., and S.D. Thompson. 1988. Energy allocation in mammalian reproduction. American Zoologist 28:863–875.

Goodpaster, W.W., and D.F. Hoffmeister. 1954. Life history of the golden mouse, *Peromyscus nuttalli*, in Kentucky. Journal of Mammalogy 35:16–27.

Grodzinski,W., and B.A. Wunder. 1975. Ecological energetics of small mammals. Pages 173–204 *in* F.B. Golley, K. Petrusewicz, and L. Ryszkowski, editors. Small mammals: Their productivity and population dynamics. Cambridge University Press, Cambridge, United Kingdom.

Ivey, R.D. 1949. Life history notes on three mice from the Florida east coast. Journal of Mammalogy. 30:157–162.

Jewell, M.A., M.K. Anderson, and G.W. Barrett. 1991. Bioenergetics of the golden mouse on experimental sumac seed diets. American Midland Naturalist 125:360–364.

Kaczmarski, F. 1966. Bioenergetics of pregnancy and lactation in the bank vole. Acta Theriologica 11:409–417.

Knuth, B.A., and G.W. Barrett. 1984. A comparative study of resource partitioning between *Ochrotomys nuttalli* and *Peromyscus leucopus*. Journal of Mammalogy 65:576–583.

Lackey, J.A. 1985. *Peromyscus leucopus*. Mammalian Species 247:967–976.

Layne, J.N., and P.G. Dolan. 1975. Thermoregulation, metabolism, and water economy in the golden mouse (*Ochrotomys nuttalli*). Comparative Biochemistry and Physiology 52A:153–163.

Linzey, D.W., and R.L. Packard. 1977. *Ochrotomys nuttalli*. Mammalian Species 75:1–6.

Liu, H., D. Wang, and Z. Wang. 2003. Energy requirements during reproduction in female Brandt's voles (*Microtus brandtii*). Journal of Mammalogy 84:1410–1416.

Mattingly, D.K., and P.A. McClure. 1985. Energy allocation during lactation in cotton rats (*Sigmodon hispidus*) on a restricted diet. Ecology 66:928–937.

McCracken, D.W. 1978. A study of utilization and partitioning of vertical space in five species of small, woodland rodents. PhD dissertation, Wake Forest University, Wake Forest, North Carolina.

McNab, B.K. 1980. Food habits, energetics, and the population biology of mammals. American Naturalist 116:106–124.

McNab, B.K. 2002. The physiological ecology of vertebrates: A view from energetics. Cornell University Press, Ithaca, New York.

McNaughton, S.J., and L.L. Wolf. 1970. Dominance and the niche in ecological systems. Science 167:131–139.

Meserve, P.L. 1977. Three-dimensional home ranges of cricetid rodents. Journal of Mammalogy 58:549–558.

Mitton, J.B. 1993. Theory and data pertinent to the relationship between heterozygosity and fitness. Pages 17–41 *in* W. Shields and N. Thornhill, editors. The natural history of inbreeding and outbreeding. University of Chicago Press, Chicago, Illinois.

Mitton, J.B., and M.C. Grant. 1984. Associations among protein heterozygosity, growth rate, and developmental homeostasis. Annual Review of Ecology and Systematics 15:479–499.

Morzillo, A.T., G.A. Feldhamer, and M.C. Nicholson. 2003. Home range and nest use of the golden mouse (*Ochrotomys nuttalli*) in southern Illinois. Journal of Mammalogy 84:553–560.

Odum, H.T., and E.C. Odum. 1976. Energy basis for man and nature. McGraw-Hill Inc., New York, New York.

O'Malley, M., J. Blesh, M. Williams, and G.W. Barrett. 2003. Food preferences and bioenergetics of the white-footed mouse (*Peromyscus leucopus*) and the golden mouse (*Ochrotomys nuttalli*). Georgia Journal of Science 61:233–237.

Peles, J.D., and J.F. Merritt. 2005. Allozyme heterozygosity and metabolic rate in *Blarina brevicauda*. Pages 367–372 in J. F. Merritt, S. Churchfield, R. Hutterer, and B. I. Sheftel, Special Publications of the International Society of Shrew Biologists. International Society of Shrew Biologists, New York, New York.

Peles, J.D., C.K. Williams, and G.W. Barrett. 1995. Bioenergetics of golden mice: The importance of food quality. American Midland Naturalist 133:373–376.

Seagle, S.W. 1985. Competition and coexistence of small mammals in an east Tennessee pine plantation. American Midland Naturalist 114:272–282.

Springer, S.D., P.A. Gregory, and G.W. Barrett. 1981. Importance of social grouping on the bioenergetics of the golden mouse, *Ochrotomys nuttalli*. Journal of Mammalogy 62:628–630.

Stueck, K.L., M.P. Farrell, and G.W. Barrett. 1977. Ecological energetics of the golden mouse based on three laboratory diets. Acta Theriologica 22:309–315.

Thompson, S.D. 1992. Gestation and lactation in small mammals: Basal metabolic rate and the limits of energy use. Pages 213–259 in T.E. Tomasi and T.H. Horton, editors. Mammalian energetics: Interdisciplinary views of metabolism and reproduction. Cornell University Press, Ithaca, New York.

Tomasi, T.E., and T.H. Horton, editors. 1992. Mammalian energetics: Interdisciplinary views of metabolism and reproduction. Cornell University Press, Ithaca, New York.

Wagner, DM., G.A. Feldhamer, and J.A. Newman. 2000. Microhabitat selection by golden mice (*Ochrotomys nuttalli*) at arboreal nest sites. American Midland Naturalist 144:220–225.

Young, B.L., and J. Stout. 1986. Effects of extra food on small rodents in a south temperate zone habitat: Demographic responses. Canadian Journal of Zoology 64:1211–1217.

9
Nesting Ecology of the Golden Mouse: An *Oikos* Engineer

Thomas M. Luhring and Gary W. Barrett

One exciting prospect for the concept of ecosystem engineering is its potential to link across levels of biological organization and approaches. (Wright and Jones 2006:207)

Most bird and small mammal species construct a single type of nest that is specific to that species' habitat needs. For example, the old-field mouse, *Peromyscus polionotus*, constructs a burrow nest where more than one exit connects the nest chamber, whereas the deer mouse, *P. maniculatus*, constructs a simple chamber at the end of a short burrow (Dawson et al. 1988, Hoffmann 1994). The golden mouse, *Ochrotomys nuttalli*, however, appears to have evolved a plastic nesting behavior that has allowed it to create a diversity of nest types.

Based on both personal observation and the literature, a diversity of golden mouse nest types exists. We describe three basic groups of nests that can be further subdivided into seven basic types (Figure 9.1). The globular arboreal nest is the most commonly encountered type of nest and is usually associated with thick undergrowth vegetation (Blus 1966, Linzey and Packard 1977, Morzillo et al. 2003, Wagner et al. 2000). This nest type seems to be used most frequently for breeding purposes and is usually inhabited by females with or without young (Morzillo et al. 2003). Morzillo et al. (2003) found males to be present near females in arboreal nests but did not find males in the nest with the females during the day. Additionally, we found no records of adult males being in a globular nest with an adult female when young were present.

Nest Types

Literature describing the nesting behavior of golden mice is robust (Table 9.1). Most articles noted the globular arboreal nest (Figure 9.2A) located from near ground level to >10 m in height (Barbour 1942, Blus 1966, Frank and Layne 1992, Goodpaster and Hoffmeister 1954, Linzey and Packard 1977, Morzillo et al. 2003). Other researchers, however, noted large communal/shelter nests

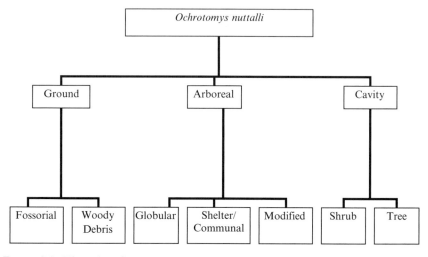

FIGURE 9.1. Hierarchy of nest types constructed by the golden mouse (*O. nuttalli*).

TABLE 9.1. Nest types of the golden mouse (*O. nuttalli*).

Nest types	Description	References
Globular	Globular nests constructed from shredded bark, grasses, and vines located in heavy undergrowth	Handley 1948; Ivey 1949; Barbour 1951; Goodpaster and Hoffmeister 1954; Packard and Garner 1964; Linzey 1968; Frank and Layne 1992; Morzillo et al. 2003; Luhring and Barrett (personal observation)
Shelter/communal	Large nests composed of dried leaves, sticks, twigs, thorns, and nest materials	Barbour 1942; Dunaway 1955; Blus 1966; Stueck et al. 1977; Morzillo et al. 2003
Modified	Modified bird or squirrel nests with nesting material added	Goodpaster and Hoffmeister 1954; Morzillo et al. 2003; Luhring and Barrett (personal observation)
Shrub cavities	Hollow cavities of shrubs, such as Chinese privet	Luhring and Barrett (personal observation)
Tree cavities	Hollow cavities of deciduous trees	Pickens 1927; Strecker and Williams 1929; Goodpaster and Hoffmeister 1954
Fossorial	Nests underground or under tree stumps	Pearson 1954; McCarley 1958; Easterla 1967; Frank and Layne 1992; Morzillo et al. 2003
Woody Debris	Nests at ground level; under or within woody debris, such as logs	Strecker and Williams 1929; Barbour 1942; Eads and Brown 1953; Pearson 1954; McCarley 1958

Figure 9.2. (A) Globular arboreal nest in Chinese privet (*L. sinense*); photograph by Thomas Luhring. (B) Shelter/communal nest in a honey locust tree (*Gleditsia triacanthos*); photograph by Terry L. Barrett.

(*Continued*)

FIGURE 9.2. cont'd. (C) A ground nest of the golden mouse located under a turkey oak (*Q. laevis*) tree; photograph by James N. Layne. (D) Golden mouse on woody debris under which a nest is located; photograph by Thomas Luhring.

(Continued)

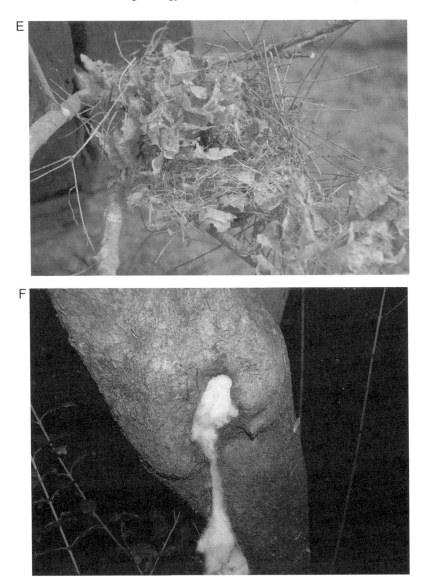

FIGURE 9.2. cont'd. (E) A modified golden mouse nest originally built by a bird; photograph by Thomas Luhring. (F) A golden mouse moved this cotton from a nearby live trap in preparation for nesting in a *L. sinense* shrub cavity; photograph by Thomas Luhring.

(Figure 9.2B) (Barbour 1942, Blus 1966, Dunaway 1955, Stueck et al. 1977), nests constructed at ground level or underground (Figure 9.2C) (Easterla 1967, Frank and Layne 1992, McCarley 1958, Morzillo et al. 2003, Pearson 1954), nests under or within woody debris (Figure 9.2D) (Strecker and Williams 1929), modified nests originally engineered by other species (Figure 9.2E) (Goodpaster and Hoffmeister 1954, Morzillo et al. 2003), and nests in shrub or tree cavities have also been reported (Figure 9.2F) (Goodpaster and Hoffmeister 1954, Pickens 1927, Strecker and Williams 1929).

Barbour (1942) suggested that golden mice construct two main types of nest: a globular nest used for rearing young and found in greenbrier (*Smilax* sp.) tangles, small pines, or deciduous bushes, and a relatively larger shelter/communal nest used for group shelter. As many as six to eight individuals have been observed in shelter/communal nests (Barbour 1942, Springer et al. 1981, Stueck et al. 1977). Stueck et al. (1977) reported finding six golden mice in a large shelter nest located in an Eastern red cedar (*Juniperus virginiana*) tree. Springer et al. (1981) reported groupings ranging from two to six individuals per nest in Madison County, Kentucky. Goodpaster and Hoffmeister (1954) also reported finding six individuals (three males and three females) in a nest in Rowan County, Kentucky.

In addition to difference in size, the globular nest appears to be transient in use, whereas the shelter/communal nest is reused by subsequent generations (Goodpaster and Hoffmeister 1954). The shelter/communal nest is used by several golden mice concurrently and likely provides an increase in energetic efficiency during the cooler winter months (Knuth and Barrett 1984, Springer et al. 1981).

Morzillo et al. (2003) found both arboreal and fossorial ground nests constructed by golden mice in southern Illinois. The role of the fossorial nest is well documented to be the dominant nest type in central Florida (Frank and Layne 1992) but its use is poorly understood and its abundance is possibly underestimated due to the difficulty in finding such a cryptic nest type (Whitaker and Hamilton 1998). Golden mouse nests made from modified squirrel and bird nests, as well as tree cavity nests have also been reported (Goodpaster and Hoffmeister 1954, Morzillo et al. 2003), but a paucity of information is available regarding their role in the ecology of the golden mouse.

Use of tree cavities by golden mice has been reported by Goodpaster and Hoffmeister (1954), Pickens (1927), and Strecker and Williams (1929). We also observed golden mice using the same type of shrub (Chinese privet, *Ligustrum sinense*) cavities as white-footed mice (*P. leucopus*) at the HorseShoe Bend Experimental Site near Athens, Georgia. Although these two species of small mammals exhibit overlap in use of habitat space (Christopher and Barrett 2006, O'Malley et al. 2003), food resource partitioning (Knuth and Barrett 1984), microhabitat utilization (Seagle 1985), nesting site selection (Morzillo et al. 2003), and the engineering ability of golden mice might reduce levels of competition between these two species if nesting sites (e.g., shrub and tree cavities) are a limiting resource. Future studies questioning how this diversity of nest types relates to levels of reproduction, predation, and survivorship would be instructive.

Nest Materials

Golden mice use a high diversity of materials in nest construction (Table 9.2), including items discarded by humans (e.g., bits of cloth, newspapers) to natural products (e.g., bird feathers, Spanish moss, dried leaves). Golden mouse globular nests typically have an outer covering composed of thatched bark, twigs, vines, and leaves, with softer inner parts including rabbit fur, finely shredded bark, grasses, palmetto fibers, bird feathers, thistle plumes, and shredded

TABLE 9.2. Nest material of the golden mouse (*O. nuttalli*).

Material	Habitat	References
Finely shredded parts of sedges and grasses; palmetto; thistle plumes; hemp fibers	Southern ridges and hill habitat (FL); upland pine-oak woodland (TX); floodplain of Mississippi River (IL); barrier island of east coast (FL)	Ivey 1949; Pearson 1954; Packard and Garner 1964; Blus 1966; Linzey 1968; Frank and Layne 1992
Shredded bark (basswood, Japanese honeysuckle, cedar, grape vine bark)	Southern ridge sandhill habitat (FL); barrier island (FL); Big Black Mountain (KY); Great Smoky Mountains (TN)	Handley 1948; Ivey 1949; Barbour 1951; Layne 1958; Packard and Garner 1964; Linzey 1968; Frank and Layne 1992
Bird feathers (cardinal, chicken, bobwhite, junco, mallard, red-eyed vireo, starling, woodcock)	Allegheny Plateau (KY); Rowan County (KY); floodplain of Mississippi River (IL); mesic hardwood forests (IL); Great Smoky Mountains (TN)	Barbour 1942; Goodpaster and Hoffmeister 1954; Layne 1958; Packard and Garner 1964; Blus 1966; Linzey 1968
Bits of newspaper; bits of cloth	Pines and hardwoods in Allegheny Plateau (KY); hardwood forests (IL)	Goodpaster and Hoffmeister 1954; Blus 1966
Rabbit fur; "down" from seed pods of milkweed plants; "cotton" from cottonwood trees; cluster of thorns of honey locust tree	Allegheny Plateau (KY); floodplain of Mississippi River (IL); floodplain of North Oconee River (GA)	Goodpaster and Hoffmeister 1954; Blus 1966; Luhring and Barrett (personal observation)
Dried leaves (catbrier, cane, box elder, honeysuckle, maple, pin oak, grape, ash, elm, sassafras)	Floodplain of Mississippi River (IL)	Ivey 1949; Goodpaster and Hoffmeister 1954; Packard and Garner 1964; Blus 1966; Linzey 1968; Frank and Layne 1992
Spanish moss	Coastal Gulf Hammock (FL); Barrier Island of Florida east coast (FL)	Ivey 1949; Pearson 1954; Frank and Layne 1992

Japanese honeysuckle bark (Frank and Layne 1992, Goodpaster and Hoffmeister 1954, Layne 1958, Linzey 1968, Packard and Garner 1964, Pearson 1954). Thus, golden mice exhibit similar plasticity in nest material selection as they do in diversity of nest types.

Concept of Ecosystem Engineer

Jones et al. (1994, 1997) and Jones and Lawton (1995) anchored the concept of species as ecosystem engineers. Alper (1998) discussed the significance of this concept, which holds that ecosystem engineers alter habitat through two mechanisms. The first mechanism is as *autogenic engineers*, which transform ecosystems by their activities and behavior to become an integral part of the altered habitat and environment. The alternative mechanism is as *allogenic engineers*, which alter the environment and then move on, leaving structures behind. We suggest a third mechanism of engineering: a species that remains an integral part of a community by altering its engineering behavior (i.e., it creates a diversity of specially modified abodes) through evolutionary time to increase its survivorship.

We hypothesize that the diversity of nest types constructed by golden mice increases species survivorship by decreasing rates of predation (O'Malley et al. 2003, Wagner et al. 2000), increasing energetic efficiency (especially during winter months) (Springer et al. 1981), increasing dietary and food quality selection (Peles et al. 1995), and reducing competition with similar small mammal species by differential use of habitat space (Christopher and Barrett 2006). The reproductive success and survivorship of golden mice throughout their geographic range is likely related to their capabilities for constructing a wide variety of specially engineered and modified abodes to fit particular circumstances. Because the Greek term *oikos* means "household," we suggest that the golden mouse functions as an "*oikos engineer*" within various ecosystem types.

Experimental Studies

Resource Manipulation

Experiments were designed to test the hypothesis that food resources (seeds or nuts) or nesting materials (cotton) were likely limiting factors affecting the population abundance of this relatively rare species. Preliminary data were also collected to determine if the addition of artificial nests might increase nesting activity and, perhaps, increase population abundance. To describe nest types, season of most frequent use, and function of each nest type in the Georgia Piedmont, we conducted systematic, seasonal surveys of golden mouse nests during autumn, winter, and spring.

Our study was conducted at the HorseShoe Bend Experimental Site (HSB), located near Athens, Georgia (33°57'N, 83°23'W) on the Georgia Piedmont. HSB is a 14.2-ha riverine peninsula formed by a meander of the North Oconee River and is located within the upper terrace of the floodplain of the Oconee River Watershed (House et al. 1984). Barrett (1968) and Hendrix (1997) described the early history and physiogeography of the HSB site, now over 40 years of secondary succession.

Our study and field observations were conducted in the upland and lowland areas of this forested peninsula. Water oak (*Quercus nigra*), sweet gum (*Liquidambar styraciflua*), and Chinese privet were the dominant woody vegetation at the study site (Klee et al. 2003). River birch (*Betula nigra*) and white oak (*Quercus alba*) were found mainly in the lowland and upland habitats, respectively. Both the upland and lowland areas contain abundant greenbrier (*Smilax* sp.) and Amur and Japanese honeysuckle (*Lonicera maackii* and *L. japonica*). These plant species provide important food, cover, and nesting material for several species of small mammals, especially for *O. nuttalli* (Christopher and Barrett 2006).

Because *O. nuttalli* is a relatively rare small mammal species (see Chapter 7 of this volume), we were interested in determining if nest sites (e.g., cavities) or nesting materials (e.g., cotton) might be limiting factors regulating population abundance. To test this hypothesis, eight experimental grids (0.21-ha each) were established during the winter of 2003–2004. Four grids were located in the lowland habitat and four in the upland habitat. Each grid contained 12 trapping stations. Each trapping station consisted of a Sherman live trap ($7.6 \times 7.6 \times 25.4$ cm) placed in a wooden shelter attached to a wooden platform on a tree at a height of 1.5 m. Stations were arranged in two parallel rows ~10 m apart within each grid (see Christopher and Barrett 2006 for details). Traps in each experimental grid were randomly assigned one of four treatments: cotton only, seed only, cotton and seed, and a control with neither cotton nor seed. Thus, three trap stations represented each treatment within each grid. Traps were checked weekly from 21 October 2003 to 8 March 2004 for nesting activity.

To investigate the importance of nesting material compared to food resources, two additional experimental grids ($n = 10$) were established on 19 September 2004. Beginning 1 October 2004, either cotton and sunflower seeds or sunflower seeds only were supplied within each grid to determine if nesting material or food might be a limiting factor regarding the reproductive success and population abundance of golden mice. Traps were observed weekly for nesting activity from 16 November 2004 to 6 March 2005. A second Sherman live trap was added to each station at ground level on 13 March 2005 (24 traps per grid or 240 traps total). Live trapping was conducted twice weekly from 30 March to 13 May 2005 to quantify the mean relative population abundance per treatment.

Our response variable was the mean number of nests per trap night for the three trapping stations belonging to each treatment within each grid. We tested the effects of experimental treatment ($n = 8$ replicates per treatment), block (lowland versus upland habitat), and their interaction on nesting activity using a two-way analysis of variance (Potvin 2001). We then used Scheffe's multiple comparison test to determine differences among treatments. Significance was established at the $p \leq 0.05$ level of probability.

Nest Searches

Systematic monthly surveys were conducted from December 2004 through March 2005 to locate arboreal nests of golden mice. Each survey was conducted by two to three experienced observers who walked along parallel transects (spaced ~5 m apart) through each experimental grid. Observers noted nest condition, materials used, presence of young or adult individuals, and vegetative habitat of each nest site. A small probe was gently entered into each nest until movement was observed or a golden mouse either exited or looked out of the entrance when disturbed (see also Wagner et al. 2000). Each nest was marked with a white utility flag to monitor nesting behavior during each successive survey.

Artificial Nests for Golden Mice

Abundance of formerly rare species, such as the Eastern bluebird (*Sialia sialis*) and the purple martin (*Progne subis*), increased when artificial nest boxes were provided to enhance reproductive success. Nest boxes have routinely been used to monitor population densities of white-footed mice (*P. leucopus*) (Lewellen and Vessey 1998, Wilder et al. 2005), a small mammal species of similar body mass and natural history as the golden mouse (Christopher and Barrett 2006). We investigated if a relatively rare species of small mammal, such as the golden mouse, would use and benefit from an artificial nest structure. To address this question, we constructed artificial nests ($n = 20$) and then situated each nest in rows of Chinese privet that lined each side of the entrance road to the HSB experimental site.

The rectangular exterior box frame of each artificial nest was constructed of 0.64-cm poplar dowel rods lashed together with sisal rope. The interior nest area within each frame was ~30 cm × 30 cm × 60 cm and filled with small dead tree branches. Dried grass and nonabsorbent cotton were added near the center as interior nest material. Artificial nests were established 1.0–2.5 m above ground level on 12 November 2005 and observed for nesting activity on 28 January, 28 February, and 28 March 2006.

Results and Observations

We sampled a total of 492 trap nights per treatment during the nesting experiment in the winter of 2003–2004. Significantly greater nesting activity of golden mice was observed in traps with cotton and seed (total of 77 nests) and in traps with cotton only (74) than in traps with seed only (41) or in controls (12). This finding suggests

TABLE 9.3. Golden mouse nesting activity from 21 October 2003 to 8 March 2004 in eight experimental trap grids at the HorseShoe Bend Ecological Research Site near Athens, Georgia.

	Cotton	Cotton and seed	Seed only	Control	
Nests	74	77	41	12	
Nests/trap night	0.15	0.16	0.08	0.02	
Source of variation	*df*	Type I SS	Mean square	*F*-Value	*p*
BLOCK	1	3.12	3.12	0.09	0.77
TRTMT	3	353.25	117.75	3.33	< 0.05
BLOCK TRTMT	3	87.62	29.21	0.83	0.49
Error	24	847.50	35.31		

that nonabsorbent cotton is a functional nesting material for golden mice. Although nesting activity was significantly different among treatments ($F_{3,24} = 3.33$, $p = 0.04$ [Table 9.3]), pairwise comparisons revealed no significant differences ($p > 0.05$) between individual treatments. Also, we found no significant difference between upland and lowland habitats ($F_{1,24} = 0.09$, $p = 0.77$) and no significant interaction between treatment and habitat ($F_{3,24} = 0.83$, $p = 0.49$ [Table 9.3]).

The addition of nesting materials in live traps increased nesting activity but did not result in a significant increase in the population abundance of golden mice. This increase in nest activity appears to reflect the selection of nest sites based on the availability of nesting materials more than food resources. The lack of a measurable effect of food resource availability on population abundance was likely the interactive result of natural resource availability (e.g., water oak acorns [Christopher and Barrett 2006]), predation, or plasticity in nesting behavior.

During the second nesting experiment conducted from 16 November 2004 to 6 March 2005, we observed moderate nesting activity ($n = 33$) in grids supplied with cotton and seeds but none in grids with seeds only. Population abundance of golden mice was compared between the two treatments and each had the same mean minimum population density of 1.8 golden mice/ha during spring 2005.

Nest surveys from 4 December 2004 to 5 March 2005 revealed 76 globular arboreal nests (4 of which were active), 20 cavity nests, 1 active communal/shelter nest, and 1 modified bird nest. Figure 9.1 summarizes the seven nest types described by other investigators. Cotton was present in all nests (except the modified bird nest), which further suggests that nonabsorbent cotton is a suitable nesting material.

Although the arboreal globular nest was found in greatest abundance at our study site during 2004–2005, few contained golden mice. One globular nest had two adult individuals on 4 December 2004 situated 1.5 m from the ground among branches of Chinese privet. During the same survey, we also observed an adult female with a juvenile in a globular nest. This juvenile was likely less than 2 weeks old, as it was hairless and its eyes had not yet opened (Linzey and Linzey 1967).

Three (15 percent) of the artificial nests that we added to the edge habitat were occupied on each date of observation. We observed two *O. nuttalli* nests and one *P. leucopus* nest on 28 January, one *O. nuttalli* nest and two *P. leucopus* nests on both 28 February and 28 March; at least one newly occupied nest was observed on each sample date. This finding adds credence to the coexistence and niche overlap of both white-footed mice and golden mice using similar habitat space and nesting resources in the same ecosystem type. The observation also suggests that the addition of artificial nests in edge habitat might enhance population abundance and reproductive success of this unique species. Future studies are planned to test this hypothesis.

Golden Mice as *Oikos* Engineers

As noted earlier, Jones et al. (1994, 1997) established the concept of the ecosystem engineer. Wright and Jones (2006), in an article entitled "Organisms as Ecosystem Engineers," summarized the concept a decade later, outlining many of the controversies surrounding the concept and describing some of the major insights gained by viewing ecological systems through the lens of ecosystem engineering. Reichman and Seabloom (2002) cautioned against overuse of this concept and suggested that the term "ecosystem engineer" be restricted to cases in which the physical modification of the environment is large relative to biotic/abiotic processes operating within the system. Most obvious examples of this concept include beavers (*Castor canadensis*), bison (*Bison bison*), and African elephants (*Loxodonta africana*), which manifest large engineering effects. A paucity of information exists, however, on how and if small mammals might function as ecosystem engineers due to their diversity of nest-type construction, the relationship of this "engineering" function to other species within the small mammal community, and whether community-level effects (e.g., rates of predation, food or seed abundance, and exploitation competition) are manifested in forest ecosystem structure and function.

We contend that the golden mouse represents a small mammal species that, because of its diversity of nest-type construction, might serve as a model species to investigate this cascading "bottom-up" impact through higher levels of ecological organization. We not only describe the diversity of nest types constructed, including the materials used in these nest types, but also conduct experiments and make observations to address this perspective. A better understanding of nesting behavior should provide insight into the evolutionary biology of this unique small mammal species.

Concluding Remarks

Much attention has focused on competition as a mechanism of population regulation (e.g., Brown et al. 1979, Giller 1984), with less attention focused on mechanisms that enhance coexistence. For example, the concepts of guilds and niche theory were developed based on competition for resources. We suggest that future investigations of coexistence and population regulation of golden mice focus on mechanisms of sociality and landscape architecture. Mechanisms of golden mouse coexistence with other small mammal species thus far include resource partitioning (Knuth and Barrett 1984), dictary diversity (Peles et al. 1995), and use of habitat space (Christopher and Barrett 2006). We suggest that the plasticity of nest-building behavior exhibited by *O. nuttalli* represents yet another evolutionary mechanism that permits coexistence with other small mammals of similar life histories and body mass, such as the white-footed mouse. We suggest that this nest-building behavior qualifies golden mice as a home-building (*oikos*) engineer.

Acknowledgments. We thank R. Adams, M. Beres, D. Crawford, T. Gancos, L. Gibbes, M. Horne, D. Howington III, B. Hughes, C. Jennison, H. Korngold, K. Meek, A. Peachy, L. Rodas, C. Schmidt, and S. Shivers for field assistance, G. Feldhamer and J. Gibbons for helpful critiques of this Chapter, and B. Rothermel for statistical advice.

Literature Cited

Alper, J. 1998. Ecosystem "engineers" shape habitats for other species. Science 280:1195–1196.

Barbour, R.W. 1942. Nests and habitat of the golden mouse in eastern Kentucky. Journal of Mammalogy 23:90–91.

Barbour, R.W. 1951. Mammals of Big Black Mountain, Harlan County, Kentucky. Journal of Mammalogy 32:546–547.

Barrett, G.W. 1968. Effects of an acute insecticide stress on a semi-enclosed grassland ecosystem. Ecology 49:1019–1035.

Blus, L.J. 1966. Some aspects of golden mouse ecology in southern Illinois. Transactions of the Illinois Academy of Sciences 59:334–341.

Brown, J.H., D.J. Reichmann, and D.W. Davidson. 1979. Granivory in desert ecosystems. Annual Review of Ecology and Systematics 10:210–227.

Christopher, C.C., and G.W. Barrett. 2006. Coexistence of white-footed mice (*Peromyscus leucopus*) and golden mice (*Ochrotomys nuttalli*) in a southeastern forest. Journal of Mammalogy 87:102–107.

Dawson, W.D., C.E. Lare, and S.S. Schumpert. 1998. Inheritance of burrow building in *Peromyscus*. Behavior Genetics 18:371–382.

Dunaway, P.B. 1955. Late fall home ranges of three golden mice, *Peromyscus nuttalli*. Journal of Mammalogy 36:297–298.

Eads, J.H., and J.S. Brown. 1953. Studies on the golden mouse, *Peromyscus nuttalli aureolus,* in Alabama. Journal of the Alabama Academy of Sciences 25:25–26.

Easterla, D.A. 1967. Terrestrial home site of the golden mouse. American Midland Naturalist 79:246–247.

Frank, P.A., and J.N. Layne. 1992. Nests and daytime refugia of cotton mice (*Peromyscus gossypinus*) and golden mice (*Ochrotomys nuttalli*) in south-central Florida. American Midland Naturalist 127:21–30.

Giller, P.S. 1984. Community structure and the niche. Chapman & Hall, London, United Kingdom.

Goodpaster, W.W., and D.F. Hoffmeister. 1954. Life history of the golden mouse, *Peromyscus nuttalli*, in Kentucky. Journal of Mammalogy 35:16–27.

Handley, C.O., Jr. 1948. Habitat of the golden mouse in Virginia. Journal of Mammalogy 29:298–299.

Hendrix, P.F. 1997. Long-term patterns of plant production and soil carbon dynamics in a Georgia agroecosystem. Pages 235–245 *in* E.A. Paul, R. Paustian, E.T. Elliott, and C.Y. Cole, editors. Soil organic matter in temperate agroecosystems. CRC Press, Boca Raton, Florida.

Hoffmann, A.A. 1994. Behavior genetics and evolution. Pages 7–42 *in* P.J.B. Slater and T.R. Halliday, editors. Behavior and evolution. Cambridge University Press, Cambridge, United Kingdom.

House, G.J., B.R. Stinner, D.A. Crossley, Jr., and E.P. Odum. 1984. Nitrogen cycling in conventional and no-tillage agro-ecosystems: Analysis of pathways and processes. Journal of Applied Ecology 21:991–1012.

Ivey, R.D. 1949. Life history notes on three mice from the Florida east coast. Journal of Mammalogy 30:157–162.

Jones, C.G., and J.H. Lawton. 1995. Linking species and ecosystems. Chapman & Hall, New York, New York.

Jones, C.G, J.H. Lawton, and M. Shachak. 1994. Organisms as ecosystem engineers. Oikos 69:373–386.

Jones, C.G, J.H. Lawton, and M. Shachak. 1997. Positive and negative effects of organisms as physical ecosystem engineers. Ecology 78:1946–1957.

Klee, R.V., A.C. Mahoney, C.C. Christopher, and G.W. Barrett. 2003. Riverine peninsulas: An experimental approach to homing in white-footed mice (*Peromyscus leucopus*). American Midland Naturalist 151:408–413.

Knuth, B.A., and G.W. Barrett. 1984. A comparative study of resource partitioning between *Ochrotomys nuttalli* and *Peromyscus leucopus*. Journal of Mammalogy 65:576–583.

Layne, J.N. 1958. Notes on mammals of southern Illinois. American Midland Naturalist 60:219–254.

Lewellen, R.H., and S.H. Vessey. 1998. The effects of density dependence and weather on population size of a polyvoltine species. Ecological Monographs 68:571–594.

Linzey, D.W. 1968. An ecological study of the golden mouse, *Ochrotomys nuttalli*, in the Great Smoky Mountains National Park. American Midland Naturalist. 79:320–345.

Linzey, D.W., and A.V. Linzey. 1967. Growth and development of the golden mouse, *Ochrotomys nuttalli nuttalli*. Journal of Mammalogy 48:445–458.

Linzey, D.W., and R.L. Packard. 1977. *Ochrotomys nuttalli*. Mammalian Species 75:1–6.

McCarley, H. 1958. Ecology, behavior and population dynamics of *Peromyscus nuttalli* in eastern Texas. Texas Journal of Science 10:147–171.

Morzillo, A.T., G.A. Feldhamer, and M.C. Nicholson. 2003. Home range and nest use of the golden mouse (*Ochrotomys nuttalli*) in southern Illinois. Journal of Mammalogy 84:553–560.

O'Malley, M., J. Blesh, M. Williams, and G.W. Barrett. 2003. Food preferences and bioenergetics of the white-footed mouse (*Peromyscus leucopus*) and the golden mouse (*Ochrotomys nuttalli*). Georgia Journal of Science 61:233–237.

Packard, R.L., and H. Garner. 1964. Arboreal nests of the golden mouse in eastern Texas. Journal of Mammalogy 45:369–374.

Pearson, P.G. 1954. Mammals of Gulf Hammock, Levy County, Florida. American Midland Naturalist 51:468–480.

Peles, J.D., C.R. Williams, and G.W. Barrett. 1995. Bioenergetics of golden mice: The importance of food quality. American Midland Naturalist 133:373–376.

Pickens, A.L. 1927. Golden mice in upper South Carolina. Journal of Mammalogy 8:246–248.

Potvin, C. 2001. ANOVA experimental layout and analysis. Pages 63–76 *in* S.M. Scheiner and J. Gurevitch, editors. Design and analysis of ecological experiments, 2nd ed. Oxford University Press, Oxford, United Kingdom.

Reichman, O.J., and E.W. Seabloom. 2002. Ecosystem engineering: A trivialized concept? Trends in Ecology and Evolution 17:308.

Seagle, S.W. 1985. Patterns of small mammal microhabitat utilization in cedar glade and deciduous forest habitats. Journal of Mammalogy 66:22–35.

Springer, S.D., P.A. Gregory, and G.W. Barrett. 1981. Importance of social grouping on bioenergetics of the golden mouse, *Ochrotomys nuttalli*. Journal of Mammalogy 62:628–630.

Strecker, J.K. and W.J. Williams. 1929. Mammal notes from Sulphur River, Bowie County, Texas. Journal of Mammalogy 10:259.

Stueck, K.L., M.P. Farrell, and G.W. Barrett. 1977. Ecological energetics of the golden mouse based on three laboratory diets. Acta Theriologica 22:309–315.

Wagner, D.M., G.A. Feldhamer, and J.A. Newman. 2000. Microhabitat selection by golden mice (*Ochrotomys nuttalli*) at arboreal nest sites. American Midland Naturalist 144:220–225.

Whitaker, J.O., Jr., and W.J. Hamilton, Jr. 1998. Mammals of the eastern United States. Cornell University Press, Ithaca, New York.

Wilder, S.M., A.M. Abtahi, and D.B. Meikle. 2005. The effects of forest fragmentation on densities of white-footed mice (*Peromyscus leucopus*) during the winter. American Midland Naturalist 153:71–79.

Wright, J.P, and C.G. Jones. 2006. The concept of organisms as ecosystem engineers ten years on: Progress, limitations, and challenges. BioScience 56:203–209.

10
Ectoparasites, Bots, and Vector-Borne Diseases Associated with the Golden Mouse

Lance A. Durden

No great and enduring volume can ever be written on the flea, though many there be who have tried it. (Melville 1851:456).

Golden mice, *Ochrotomys nuttalli*, are integral ecological components of woodlands in the southeastern United States and some adjacent regions, as suggested by the chapters in this volume and by publications describing the natural history and ecological relationships of this small mammal species (e.g., Christopher and Barrett 2006, Goodpaster and Hoffmeister 1954, Linzey 1968, Linzey and Packard 1977, McCarley 1958, Morzillo et al. 2003). Ecological relationships of golden mice extend to the symbiotes, including their parasitic arthropods (ectoparasites and the larvae of botflies), other epifaunistic arthropods, and microorganisms that may be transmitted by certain ectoparasites. Microorganisms could either be pathogenic or nonpathogenic to golden mice. In either case, microorganisms could cause pathology in humans resulting in zoonotic vector-borne diseases. Under these relationships, golden mice would be the *reservoir hosts* and arthropods, such as ticks, would be the *zoonotic vectors*. In this chapter, an ecological approach will be taken with respect to analyzing the ectoparasites (and phoretic arthropods) that inhabit the fur of golden mice, the bots that cause subcutaneous parasitism in this rodent, and, finally, the relationship between golden mouse ectoparasites and vector-borne diseases, including those that cause clinical disease in humans (i.e., zoonoses).

Ectoparasite Fauna of the Golden Mouse

A taxonomic listing of the 25 species of ectoparasites (and other epifaunistic arthropods) found on golden mice in the southeastern United States is given in Table 10.1. A few other less specific ectoparasite records also exist for golden mice, such as "ticks," "red bugs" (chiggers), "*Labidophorus* sp." (probably now treated as *Glycyphagus hypudaei*), and a few free-living mites (e.g., soil mites) (Linzey 1968, Linzey and Packard 1977, Pearson 1954).

TABLE 10.1. Ectoparasites recorded from the golden mouse.

Ectoparasites[a] (states)	References
Sucking lice (Phthiraptera: Anoplura)	
Hoplopleura hesperomydis (FL, GA, NC, TN)	Linzey (1968), Kim et al. (1986), Durden and Musser (1994a, 1994b), Durden et al. (1997, 2000, 2004), Ritzi (2002), Reeves et al. (2006)
Fleas (Siphonaptera)	
Ctenophthalmus pseudagyrtes (TN)	Linzey (1968), Durden and Kollars (1997), Reeves et al. (2006)
Doratopsylla blarinae (TN)	Linzey (1968), Durden and Kollars (1997), Reeves et al. (2006)
Epitedia wenmanni (GA, IL, TN)	Pfitzer (1950), Layne (1958), Linzey (1968), Durden and Kollars (1997), Hu et al. (2000), Durden et al. (2004), Reeves et al. (2006), Durden (Unpublished data)
Orchopeas leucopus (IL, KY, NC, SC, TN)	Goodpaster and Hoffmeister (1954), Layne (1958), Linzey (1968), Durden and Kollars (1997), Ritzi (2002)
Peromyscopsylla scotti (AL, GA, MS, SC)	Sanford and Hays (1974), Durden et al. (1999, 2004), Hu et al. (2000), Clark and Durden (2002)
Ticks (Acari: Ixodoidea)	
Dermacentor variabilis (FL, GA, MO, MS, NC, TN)	Morlan (1952), Linzey (1968), Durden and Kollars (1992), Durden et al. (2000, 2004), Hu et al. (2000), Kollars et al. (2000), Clark and Durden (2002), Ritzi (2002), Reeves et al. (2006)
Haemaphysalis leporispalustris (locality not listed)	Bishopp and Trembley (1945)
Ixodes cookei (locality not listed)	Bishopp and Trembley (1945)
Ixodes minor (FL, GA)	Durden et al. (2000, 2004)
Ixodes scapularis (FL)	Durden et al. (2000)
Mesostigmatid mites (Acari: Mesostigmata)	
Androlaelaps casalis (GA, TN) (syn: *Atricholaelaps megaventralis*)	Linzey (1968), Whitaker and Wilson (1974), Durden et al. (2004), Reeves et al. (2006)
Androlaelaps fahrenholzi (AL, GA, NC, TN) (syn: *Haemolaelaps glasgowi*)	Strandtmann (1949), Hays and Guyton (1958), Linzey (1968), Whitaker and Wilson (1974), Whitaker et al. (1975), Ritzi (2002), Durden et al. (2004), Reeves et al. (2006)
Eulaelaps stabularis (AL, GA, NC, TN)	Hays and Guyton (1958), Linzey (1968), Whitaker and Wilson (1974), Ritzi (2002), Durden et al. (2004), Reeves et al. (2006)
Haemogamasus liponyssoides (GA)	Durden et al. (2004)
Hypoaspis sp. (NC)	Whitaker, et al. (1975)
Laelaps alaskensis (TN)	Linzey and Linzey (1968), Reeves et al. (2006)
Laelaps nuttalli (locality not listed)	Linzey (1968), Whitaker and Wilson (1974), Linzey and Packard (1977)
Ornithonyssus bacoti (AL, GA, NC, TN) (syn: *Bdellonyssus bacoti*)	Morlan (1952), Hays and Guyton (1958), Linzey (1968), Whitaker and Wilson (1974), Ritzi (2002)

(Continued)

TABLE 10.1. Ectoparasites recorded from the golden mouse—cont'd.

Ectoparasites[a] (states)	References
Fur mites (Acari: Glycyphagidae, Myobiidae, Myocoptidae)	
Glycyphagus hypudaei (FL, GA, KY, NC, TN[b])	Linzey (1968), Fain and Whitaker (1973), Whitaker and Wilson (1974), Whitaker et al. (1975), Linzey and Packard (1977), Ritzi (2002), Durden et al. (2004), Nims et al. (2004)
Radfordia subuliger (NC)	Ewing (1938), Whitaker and Wilson (1974)
Myocoptes musculinus (NC, TN)	Linzey (1968), Whitaker and Wilson (1974), Ritzi (2002), Reeves et al. (2006)
Pygmephorid mites (Acari: Pygmephoridae)	
Bakerdania equisetosa (NC)	Ritzi (2002)
Chiggers (Acari: Trombiculidae)	
Euschoengastia peromysci (FL, GA, NC)	Farrell (1956), Durden et al. (2000, 2004)
Euschoengastia rubra (NC)	Farrell (1956)

[a]Junior synonyms are listed for species that were reported under different names in some of the citations.
[b]Reported as *Labidophorus* sp. by Linzey (1968), and Linzey and Packard (1977); however, Whitaker et al. (1975) and the current author believe this mite was *G. hypudaei*.

Only two relatively detailed studies of the ectoparasites of golden mice have been completed: one in the Great Smoky Mountains of eastern Tennessee (Linzey 1968) and the other in the Lower Coastal Plain of Georgia (Durden et al. 2004). However, less detailed records of golden mouse ectoparasites have been reported, such as those by Clark and Durden (2002), Durden et al. (2000), Hu et al. (2000), Linzey and Linzey (1968), Linzey and Packard (1977), Morlan (1952), Reeves et al. (2006), and Ritzi (2002). In addition, particular groups of ectoparasites collected from golden mice have been reported in more encompassing works on sucking lice (Durden and Musser 1994a, 1994b, Durden et al. 1997, Kim et al. 1986), fleas (Durden and Kollars 1997, Durden et al. 1999, Layne 1958, Pfitzer 1950, Sanford and Hays 1974), ticks (Bishopp and Trembley 1945, Durden and Kollars 1992), chiggers (Farrell 1956), and other mites (Ewing 1938, Fain and Whitaker 1973, Hays and Guyton 1958, Nims et al. 2004, Strandtmann 1949, Whitaker and Wilson 1974).

Compared to the number of ectoparasite species that have been recorded from certain sympatric rodent species such as the cotton mouse (*Peromyscus gossypinus*), golden mice are parasitized by a relatively low number of ectoparasite species. Typically, the numbers of ectoparasites on field-sampled golden mice are also small (Durden et al. 2004). For example, Pearson (1954) noted fewer ectoparasites on golden mice than on sympatric cotton mice. He only recorded "red bugs" (= larval chiggers) on 2 of 23 individuals (representing 94 golden mouse captures). Further, Layne (1971) reported no fleas found on 72 golden mice examined in Florida. Kollars et al. (1997) recorded no lice or fleas from 24 golden mice in Missouri. Durden et al. (2000) found no fleas or mesostigmatid mites on nine golden mice in Florida, and Clark and Durden (2002) recorded no mites (and only one species of flea and one species of tick) from six golden mice collected in Mississippi.

Similarly, Durden et al. (2004) found low numbers of 12 species of ectoparasites (and phoretic arthropods) collected from 46 golden mice in Coastal Plain Georgia.

Additional documentation of the relatively low number of ectoparasite and related arthropod species associated with golden mice comes from tallying the number of arthropod species reported from taxonomically related small mammals that are sympatric with *O. nuttalli* in different parts of its range. For example, throughout their respective ranges, at least 240 species have been reliably reported from the deer mouse (*Peromyscus maniculatus*) (2 sucking lice, 96 fleas, 15 ticks, 58 chiggers, and 69 mites belonging to other groups), 89 from the white-footed mouse (*P. leucopus*) (2 sucking lice, 28 fleas, 5 ticks, 33 chiggers, and 21 other mites), 27 from the cotton mouse (1 sucking louse, 10 fleas, 3 ticks, 4 chiggers, 9 other mites), 21 from the old-field mouse (*P. polionotus*) (1 sucking louse, 5 fleas, 2 ticks, 2 chiggers, 11 other mites), and 16 from the Florida mouse (*Podomys floridensis*) (1 sucking louse, 4 fleas, 4 ticks, 3 chiggers, and 4 other mites) (Durden et al. 2004, Layne 1963, Nims et al. 2007, Whitaker 1968, Durden Unpublished data). Except for old-field and Florida mice, more species of arthropods have been recorded from all of these rodents than from golden mice. The restricted geographical and habitat distributions of old-field and Florida mice, together with the low numbers of individuals examined for ectoparasites have probably influenced these numbers.

Several factors presumably are responsible for the relatively low level of ectoparasitism typically recorded for golden mice. Possibilities include efficient pelage grooming (self- or allo-grooming), a competent immune system, physical barriers to ectoparasite attachment, or one or more desirable ecological traits of golden mice such as use of habitat space and small home range size. Because golden mice are morphologically similar to other sympatric rodents, especially *Peromyscus* spp., for the first three of these traits, it appears likely that an ecological or behavioral trait accounts for this difference. Durden et al. (2004), for example, found lower ectoparasite species diversity and lower numbers of ectoparasites on golden mice than on sympatric cotton mice in Coastal Plain Georgia. They suggest that this difference was partly a result of the more arboreal habits of *O. nuttalli* compared to *P. gossypinus*. Representatives of the host-seeking stages of at least two major groups of ectoparasites—ticks and chiggers—occur mainly in the leaf litter and/or low vegetation (Durden et al. 2004, Sonenshine et al. 2002). Any reduction in the amount of time a potential small mammal host spends in this habitat should be beneficial with respect to a lowered incidence of exposure to questing ticks and chiggers. Thus, the tendencies of golden mice—"living on the second floor"—might help to reduce ectoparasites.

Another trait of golden mice that is probably responsible for fewer species and numbers of ectoparasites is their tendency to live in aboveground nests (Jennison et al. 2006, Morzillio et al. 2003). Rodent nests often serve as the foci for several groups of ectoparasites such as fleas and mesostigmatid mites that feed intermittently on the host (Durden and Traub 2002, Radovsky 1985). Having an arboreal nest might make such a structure difficult to colonize for certain species of ectoparasites.

Also, arboreal nest conditions, such as reduced humidity and porousness (some ectoparasite stages could fall from the nest and not be able to return easily) might be detrimental for certain arthropods. Golden mice also occupy smaller home ranges and have lower levels of activity than white-footed mice (Christopher and Barrett 2006, Goodpaster and Hoffmeister 1954, Jennison et al. 2006, Morzillo et al. 2003). The combination of these traits appears to be adaptive for golden mice in reducing their ectoparasite loads compared to sympatric rodents of a similar size, especially *Peromyscus* spp., such as white-footed mice and cotton mice.

Except for the relatively low species diversity, the ectoparasite fauna recorded for golden mice (Table 10.1) is fairly typical for a small woodland rodent native to the eastern United States. Interestingly, none of the species listed in Table 10.1 are host-specific to the golden mouse. Instead, the arthropod species listed also occur on related cricetids such as *Peromyscus* spp. (the sucking louse *Hoplopleura hesperomydis* [Figure 10.1], the fleas *Orchopeas leucopus* and *Peromyscopsylla scotti* [Figure 10.1], and the fur mite *Radfordia subuliger*) and on a variety of other rodents (the flea *Epitedia wenmanni*, immature stages of the American dog tick *Dermacentor variabilis* [Figure 10.2], the laelapid mites *Androlaelaps casalis*, *Eulaelaps stabularis*, and *Laelaps nuttalli*, the tropical rat mite *Ornithonyssus bacoti*, the fur mites *G. hypudaei* [Figure 10.3] and *Myocoptes musculinus*, and the chiggers *Euschoengastia peromysci* [Figure 10.3] and *E. rubra*). They also occur on a variety of other small mammals (the flea *Ctenophthalmus pseudagyrtes*, the laelapid mite *Haemogamasus liponyssoides*, and the pygmephorid mite *Bakerdania equisetosa*), as well as on other mammals (the laelapid mites

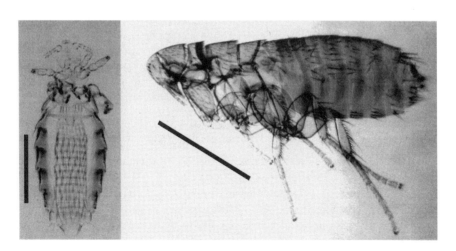

FIGURE 10.1. Left: The sucking louse, *Hoplopleura hesperomydis* (female), slide preparation, ventral view (scale bar, 500 μm). Right: The flea, *Peromyscopsylla scotti* (female), lateral view (scale bar, 1 μm).

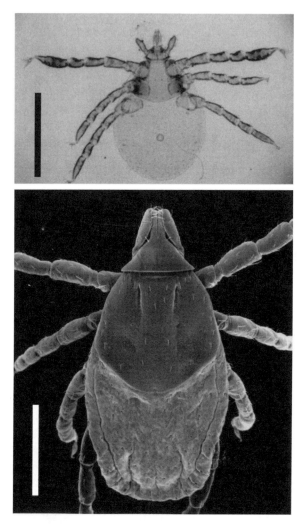

FIGURE 10.2. The tick, *Dermacentor variabilis*. Top: Larva, slide preparation, ventral view (scale bar, 500 μm). Bottom: Nymph, scanning electron micrograph, dorsal view (scale bar, 500 μm).

Androlaelaps fahrenholzi [Figure 10.3] and *Hypoaspis* spp.). These arthropods can also infest various vertebrates (immature stages of the blacklegged tick *Ixodes scapularis* which parasitize mammals, birds or reptiles), birds and rodents (the tick *Ixodes minor*), rodents and carnivores (immature stages of the carnivore tick *Ixodes cookei*), mainly shrews (the flea *Doratopsylla blarinae*), and birds and lagomorphs (immature stages of the rabbit tick *Haemaphysalis leporispalustris*). Of these 25 arthropod species, 10 of them (*H. hesperomydis,*

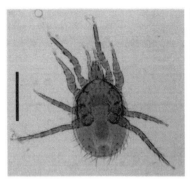 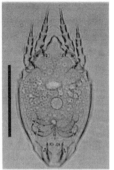 

FIGURE 10.3. Left: The mesostigmatid mite, *Androlaelaps fahrenholzi* (female), slide preparation, ventral view (scale bar, 500 μm). Middle: The fur mite, *Glycyphagus hypudaei* (phoretic deutonymphal hypopus), slide preparation, ventral view (scale bar, 200 μm). Right: The chigger, *Euschoengastia peromysci* (larva), slide preparation, ventral view (scale bar, 200 μm); note stylostome.

E. wenmanni, O. leucopus, P. scotti, D. variabilis, A. fahrenholzi, E. stabularis, O. bacoti, G. hypudaei, and *E. peromysci*) appear to be relatively common and widespread associates of the golden mouse (Table 10.1).

Life Cycles

The ectoparasites and other arthropods occurring on the golden mouse encompass a variety of life cycles. Only one of the ectoparasitic insects, the sucking louse *H. hesperomydis* (Figure 10.1), spends its entire life cycle on the host. Its legs terminate in tibiotarsal claws that are highly adaptive for clinging to host hairs (Durden 2002). Louse eggs are glued onto individual host hairs (often on the area just behind the host head on golden mice) by gravid female lice. After about a week inside the egg, the first instar nymphal lice push the cap (operculum) off the top of the egg, crawl out, await partial hardening of their exoskeleton, and then pierce the host skin to begin feeding on blood. After a few days of feeding, the first instar nymph molts into a second instar, which, in turn, molts into a third instar, each of which feeds for a few days before molting. Fully fed third instar nymphs molt into males or females, each living for up to a month feeding, mating, and (females) ovipositing on the host (Durden 2002).

Like most fleas, all flea species recorded from golden mice are only hematophagous as adults. Eggs are deposited by gravid female fleas in the host nest. The legless, eyeless flea larvae feed voraciously on various kinds of organic matter (Durden and Traub 2002). In addition, many adult fleas occur on their hosts for relatively short periods, mainly during blood-feeding, and spend the majority of their time in the host nest (Durden and Traub 2002). However, Durden et al. (2000) recovered no fleas from three arboreal golden mouse nests examined in Florida. As noted previously, arboreal nests might be less conducive to flea survival than nonarboreal rodent nests.

Like most rodents, golden mice are mainly parasitized by immature stages (larvae and nymphs) of ticks, with adult ticks typically parasitizing larger hosts (Durden 2006). However, there is a possible exception to this trend concerning golden mice. Adults of *I. minor*, a tick that in North America is confined to coastal and subcoastal regions of Florida, Georgia, and South Carolina, have been recorded from rodents collected in the southeastern United States, albeit larger species such as the cotton rat (*Sigmodon hispidus*) and eastern woodrat (*Neotoma floridana*) (Clark et al. 2001). Notwithstanding this phenomenon, immatures, especially larvae, make up the vast majority of tick records from golden mice. Occasionally, golden mice have relatively high burdens of immature *D. variabilis* (Figure 10.2) (Durden et al. 2000, 2004). As with most other similarly sized rodents, tick larvae typically attach to the pinnae of golden mice, where they have easy access to an abundant supply of blood and are not subjected to oral grooming by their hosts. Nymphal ticks might also attach to the host ears, but often they attach behind the head (between the scapulae) where host oral grooming is also avoided. Whereas *D. variabilis* (Figure 10.2) is the most frequently encountered tick on golden mice (Table 10.1), other ticks such as *I. minor* and *I. scapularis* have been reported (Durden et al. 2000, 2004). This might be significant because all three of these species of ticks are important in the maintenance or transmission of certain vector-borne disease agents, which will be discussed later. The other two tick species reported from golden mice (Table 10.1) represent atypical host associations; these ticks usually parasitize lagomorphs and birds (*H. leporispalustris*) or carnivores (*I. cookei*), respectively.

The mesostigmatid mites recorded from golden mice (Table 10.1) represent a diverse assemblage of epifaunistic arthropods with a variety of feeding strategies. All but one of these mesostigmatids belong to the family Laelapidae. The exception, *O. bacoti*, belongs to the family Macronyssidae. Radovsky (1985) detailed the evolution of host associations, life cycles, and feeding strategies of mesostigmatids associated with mammals. These mites are largely nidicolous (but also occur on the host, especially if they are lymph/blood feeders) and typically have nonfeeding larval and deutonymphal stages. The feeding protonymphal and adult female stages are usually recorded on hosts; males are more or less confined to the host nest. On-host sex ratios are female biased (sometimes in a ratio exceeding 100:1) and females of some species are also known to be parthenogenetic (arrhenotokous) (Radovsky 1985). The least derived of the laelapids, especially with respect to feeding habits, recorded from golden mice is *Hypoaspis* sp. Certain members of this genus are strongly associated with mammal nests, but sometimes they occur on the host; they are not known to feed on the host and are, therefore, phoretic rather than parasitic. In addition to being recorded from golden mice, both species of *Androlaelaps* listed in Table 10.1 have been collected from golden mouse nests (Durden et al. 2000). Species of *Androlaelaps* (Figure. 10.3), *Eulaelaps*, and *Laelaps* (Table 10.1) are all facultatively parasitic (feeding mainly on host lymph, sometimes on blood when they are parasitic). Their generalized mouthparts also allow them to imbibe fluids from host abrasions or from other arthropods (alive or dead), some of which might be ectoparasites such as sucking lice (Radovsky 1985). As an exception to these broad laelapid feeding habits, *H. liponyssoides* is hematophagous on rodent hosts (Furman 1959). The single species of macronyssid mite recorded from golden mice, *O. bacoti* (Table 10.1), is exclusively hematophagous and has highly adapted mouthparts for this specialized feeding strategy (Radovsky 1985).

The three species of fur mites recorded from golden mice are each assigned to a different family. *Glycyphagus hypudaei* (Figure 10.3) belongs to the family Glycyphagidae, members of which are characterized by having a tiny, nonfeeding "hypopial" deutonymphal stage that phoretically attaches to mammal hair using ventral claspers (Fain and Whitaker 1973, Krantz 1978). Other instars of these mites occur in mammal nests, where they are saprophages or fungivores (Krantz 1978). The phoretic deutonymphs therefore utilize their hosts as a transport mechanism for dispersal to different mammal nests. Occasionally, these mites are so common on golden mice and other rodents that they look like whitish specks dusted throughout the fur (Nims et al. 2004).

Like the sucking louse, *H. hesperomydis*, the other two small fur mites recorded from golden mice—*M. musculinus* (family Myocoptidae) and *R. subuliger* (family Myobiidae)—spend their entire life cycle on the host. Both have highly modified legs for clinging to individual host hairs. In myocoptids, legs III and IV in females and legs III in males are adapted for grasping hair, whereas in myobiids, the first pair of legs is modified for this purpose (Fain and Hyland 1985, Krantz 1978, Nutting 1985). Both myocoptids and myobiids typically feed at the bases of host hair follicles. They imbibe extracellular fluids and various secretions, although myobiids, especially gravid females, will also feed on blood (Krantz 1978). When present in large numbers, members of both of these families can cause dermatitis on their hosts (Fain and Hyland 1985, Krantz 1978, Nutting 1985).

Bakerdania equisetosa is the only pygmephorid mite that has been recorded from golden mice (Table 10.1) (Ritzi 2002). The Pygmephoridae is a fairly large assemblage of small mites that subsist on organic matter in a variety of habitats such as soil, leaf litter, plant surfaces, on insects, and in bird and mammal nests (Krantz 1978). However, members of the genus *Bakerdania* occur in the pelage of small mammals, where they are phoretic and use their host to gain access to organic food sources in host nests (Krantz 1978).

Although two species of chiggers have been recorded from golden mice (Table 10.1), only one of these, *E. peromysci* (Figure 10.3), is relatively widespread and common on *O. nuttalli*. Like other chiggers, the tiny larvae of *E. peromysci* quest for hosts from vegetation and attach to them via an elongated feeding tube called a stylostome that consists of a mixture of chigger and host components (Krantz 1978). Through the stylostome (Figure 10.3), each chigger releases saliva and enzymes and then ingests a mixture of fluids and digested host cellular material, but not blood (Krantz 1978). Engorged chiggers detach from their host but leave the stylostome in place, which causes itching, although evidently much less so in rodents than in humans. Postlarval stages (nymphs and adults) of chiggers are not parasitic; instead, they are predators on small arthropods or their eggs in soil and leaf litter (Farrell 1956, Krantz 1978). On most small North American rodents, chiggers can be common at certain times of the year. Aggregations of chigger larvae might appear as orange "patches" on the skin. In golden mice, these chigger aggregations often occur on the ears, sometimes partially inside the ear canal (personal observation).

Durden et al. (2004) found that cotton mice were parasitized by more species of arthropods (19) than were sympatric golden mice (12) in Coastal Georgia. With respect to infestation prevalences (the percentage of mice infested), four species of ectoparasites (one flea, one fur mite, and two laelapid mites) infested a significantly greater proportion of golden mice than cotton mice. One species (the tropical rat mite, O. bacoti) infested significantly more cotton mice than golden mice. Perhaps conditions of the host pelage or conditions inside the host nest affected these differences. Conversely, the nonarboreal nests typically constructed by cotton mice likely provide more conducive microhabitats for the survival and reproduction of tropical rat mites. Further, two species of ectoparasites (one flea and the blacklegged tick, I. scapularis) had significantly higher mean intensities (mean number per infested mouse) on cotton mice. However, one fur mite and one laelapid mite exhibited significantly higher mean intensities on golden mice. Again, microhabitats inside the host nests might have influenced off-host survival and reproduction for the flea, tick, and laelapid mite. Like most other ixodid ticks, I. scapularis typically quests for hosts from vegetation or leaf litter (Durden 2006). Therefore, the larger home ranges and greater activity profiles, as well as the proportion of time spent by cotton mice on the ground compared to golden mice, suggests that cotton mice should be parasitized by greater numbers of ixodid ticks.

It is interesting to compare the ectoparasites and related arthropods recorded from golden mice in the two in-depth studies on this topic: one in the higher elevations of eastern Tennessee within the Great Smoky Mountains National Park (Linzey 1968) and the other in the Lower Coastal Plain of southeastern Georgia (Durden et al. 2004). The sucking louse (H. hesperomydis), one species of flea (E. wenmanni), one species of tick (D. variabilis), three species of laelapid mites (A. casalis, A. fahrenholzi, and E. stabularis), and one species of glycyphagid fur mite (G. hypudaei) – assuming that Labidophorous sp. reported in the Tennessee survey was actually this species) were recorded in both studies. All of these ectoparasites are widespread species so it is not surprising that they were found on golden mice in both regions. However, three species of fleas (C. pseudagrytes, D. blarinae, and O. leucopus) and one species of fur mite (M. musculinus) were only recorded in the Great Smoky Mountains. One species of flea (*P. scotti*), one species of tick (*I. minor*), one species of laelapid (*H. liponyssoides*), and one species of chigger (*E. peromysci*) were only recorded in the Lower Coastal Plain of Georgia. However, four of the species that were only recorded at one of the two sites (*C. pseudagyrtes, O. leucopus, H. liponyssoides,* and *E. peromysci*) are widespread species. Likely, with more intensive ectoparasite collections from golden mice, these species would be expected to occur at both sites. The other species that were only recorded at one of the sites have more restricted distributions. For example, in North America, *I. minor* is found only in the coastal southeastern United States, *D. blarinae* not as far south as southeastern Georgia, and *P. scotti* (Figure 10.1) mainly in the southern United States but not at high elevations (Benton 1980). The other species typically parasitize other hosts—*D. blarinae* on short-tailed shrews (*Blarina brevicauda*) and *M. musculinus* on the house mouse (*Mus musculus*). Clearly, many ectoparasite species and related arthropods are widespread associates of golden mice, whereas a few species are more restricted within the geographical and elevational distribution of *O. nuttalli*.

Parasitism of Golden Mice by Botfly Larvae

Although three "species" of botfly larvae (bots) have been reported to parasitize golden mice—*Cuterebra fontinella, C. angustifrons,* and *C. grisea*—Sabrosky (1986) showed that the latter two taxa are both junior synonyms of *C. fontinella*. However, Sabrosky (1986) recognized two subspecies of *C. fontinella*. The nominate subspecies *C. f. fontinella* mainly occurs in the southeastern United States and *C. f. grisea* in western North America and parts of the northeastern United States and eastern Canada. Therefore, *C. f. fontinella* is the bot that parasitizes *O. nuttalli* throughout its range, with *C. f. grisea* occurring just outside of the geographical range of the golden mouse.

Records of botfly larvae parasitizing golden mice are available from Florida (Pearson 1954), Georgia (Durden et al. 2004, Jennison et al. 2006), Mississippi (Clark and Durden 2002), Missouri (Kollars and Durden Unpublished data), and Tennessee (Dunaway et al. 1967, Linzey 1968, Linzey and Linzey 1968). It is assumed that all of these bot larvae were *C. f. fontinella*. Currently, however, only adult botflies of the genus *Cuterebra* can be reliably identified to species or subspecies (Sabrosky 1986) and larvae were not reared to adults for determination in most of these studies. However, in the near future, molecular techniques will most likely become available for the accurate identification of certain *Cuterebra* spp. larvae (Noël et al. 2004).

The life cycle of cuterebrine botflies (Family Oestridae, subfamily Cuterebrinae) is intriguing and complex and the general characteristics for members of this subfamily have been discussed by Catts (1982). Specifically for *C. fontinella*, adult males form aggregation sites where they patrol for approaching females and, if successful, mate with them for a few minutes (Shiffer 1983).

Mated, gravid female flies later oviposit on vegetation, especially around rodent burrows and runways. Eggs are stimulated to hatch rapidly (from body warmth) when a rodent brushes against them or consumes them (Catts 1982, Durden 1995). The tiny first instar bot larvae then penetrate mucosal membranes, especially around the host mouth, nose, or eyes, and begin a remarkable journey of a few days through the body. For *C. fontinella*, this occurs mainly along the top of the head, through the dorsal midline down to the tail, and then antero-ventrally to the inguinal region of the host to a preferred subcutaneous site (Hunter and Webster 1973). During this migration stage, the bot larvae are so small that they cause little or no noticeable pathology in the host (Cogley 1991, Dunaway et al. 1967, Payne and Cosgrove 1966). For *C. fontinella*, the predilection site is almost always in the inguinal or genital region with occasional records from adjacent host body areas (Cogley 1991, Dunaway et al. 1967, Durden 1995, Gingrich 1979, Sillman and Smith 1959, Wecker 1962). When the inguinal area is reached, first instar larvae molt into second instars, start growing rapidly, and soon molt into larger third instar larvae (Figure 10.4) (Cogley 1991, Hunter and Webster 1973, Sillman and Smith 1959). Initially, the growing second instar appears as a characteristic "lump" beneath the skin. The growing larva soon requires larger amounts of oxygen than it can procure from the host tissues and forms a hole called a warble pore (Figure 10.4) to the external surface through which it respires through its posterior spiracles (Cogley 1991). Fully grown third instar

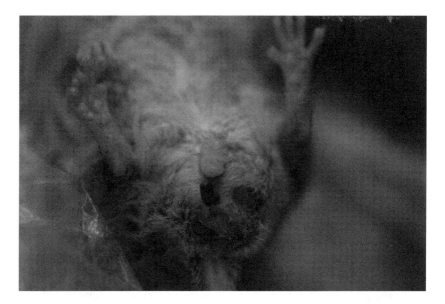

FIGURE 10.4. Male golden mouse with a single mature bot in the left inguinal region. Photograph, courtesy of Thomas Luhring, Savannah River Ecology Laboratory, Aiken, South Carolina.

botfly larvae ("warbles") can be ~20 percent of the head and body length of a golden mouse. The testes of male hosts might be displaced, or partially or completely ascended into the abdominal cavity, in the presence of mature bots. The testes descend back into the scrota, however, when the fully developed larvae leave their host (Dunaway et al. 1967). Based on studies of mature *C. fontinella* bots infesting white-footed mice, this does not adversely affect the reproductive abilities of male hosts (Catts 1982, Timm and Cook 1979). Mature, third instar larvae exit the host by crawling backwards out of the warble pore; they then burrow into the soil and pupate about 10 cm beneath the surface (Durden Unpublished data). Adult flies later (either the same year or after the winter) hatch from these pupae, crawl up through the soil, expand and harden their wings, and then repeat the life cycle. Based on experimental laboratory infestations in deer mice, the entire in-host warble development period for *C. fontinella* is 3.5–4 weeks (Cogley 1991). It is important to note that almost all studies of botfly parasitism have only recorded the easily located, larger second and third instar larvae (Figure 10.4) after they have reached their predilection sites. Because the small first instar (and sometimes also the early second instar) larvae are typically not observed and recorded, the reported intensities and prevalences of larval botfly parasitism are usually underestimated. However, scars from previous late-stage bot larvae typically remain evident on the host for a few weeks, so cases of previous infestation can frequently be noted (Catts 1982).

Although first instar *C. fontinella* larvae cause little apparent tissue damage in their hosts, the large subcutaneous cavities formed by late instar bots cause obvious tissue scarring, displacement, and reorganization. Nevertheless, healing of bot cavities caused by these larvae in white-footed mice was almost complete 9 days after bots had exited, with chronic abscesses or granulomas forming in very few hosts (Payne and Cosgrove 1966). In bot-infested white-footed mice, erythrocyte counts, hematocrit percentage, and hemoglobin counts were all lower and total leukocyte counts and fibrinogen titers were higher than in noninfested control mice (Dunaway et al. 1967). Clearly, the blood of infested hosts is dramatically affected by the presence of bot larvae. The decreased red blood cell counts could lead to other consequences such as depleted tissue oxygen, decreased host energy resources, and reduced reproductive potential of female mice. However, despite the elevated leukocyte counts during bot infestation, these parasites can obviously successfully resist or evade the cellular and humoral immunological responses mounted by the host. At least some bot species release immunosuppressive or immunoevasive chemicals that counteract immune responses mounted by the host (Baron and Colwell 1991).

Typically, individual golden mice have one late instar bot larva at a time. However, there are exceptions. Linzey (1968) recorded three inguinal bots in one golden mouse and another individual with one bot but with a scar indicating that a second bot larva had recently exited. Also, Dunaway et al. (1967) recorded two mature bot larvae in each of two golden mice and a third individual with one bot and a scar indicating a second recent bot. In southern Illinois,

74 of 1110 white-footed mice examined had bots; 61 of these mice had one bot, 11 had two bots, and 2 had three bots, with many of the animals with one bot also having fresh scars from recently emerged bots (Feldhamer Unpublished data). It would seem that a rodent as small as a golden mouse would be significantly burdened by a parasite as large as a mature botfly larva, and infestations by more than one mature larva at the same time would appear to be especially detrimental. In this respect, Dunaway et al. (1967) noted that the activity of bot-infested white-footed mice and golden mice did not seem to be greatly affected, but that some interference with locomotion was present, especially in individual mice that had more than one late instar bot at the same time. These researchers also noted that fast entry into a burrow or nest could be impeded by the greater posterior girth of mice with large inguinal warbles and that this could lead to easier capture by predators. In fact, in a laboratory setting, Smith (1978b) reported increased predation by short-tail weasels (*Mustela erminea*) on deer mice that had two or more warbles compared to deer mice with either one or no warbles. He noted that this was mainly due to the inability of infested mice to use arboreal pathways. Similarly, Smith (1978a) reported a decrease in the amount of strenuous activities, such as running in an exercise wheel or stereotypic somersaulting, in bot-infested deer mice compared to noninfested control mice. Presumably, infested wild golden mice are affected in similar ways by bot larvae.

Reported prevalences (the percentages of mice with visible bots) for botfly larval infestations are relatively low for golden mice. Dunaway et al. (1967) recorded a maximum prevalence of 1.5 percent (3 of > 200 golden mouse captures) during a 7-year field study compared to a prevalence of 24.7 percent in sympatric white-footed mice (126 of 511 captures) in Tennessee. Pearson (1954) recorded a prevalence of 8.7 percent (2 of 23 mice) in golden mice and a prevalence of 12.3 percent (33 of 269 mice) in sympatric cotton mice in Florida. Durden et al. (2004) reported comparable figures of 4.4 percent (2 of 46 golden mice) and 4.0 percent (4 of 202 cotton mice) in the Coastal Plain of Georgia. Jennison et al. (2006) recorded larval bot prevalences of 6.3–12.5 percent in golden mice and 41.7 percent in sympatric white-footed mice in the Piedmont of Georgia. Finally, in a small sample from Mississippi, Clark and Durden (2002) recorded 17 percent (1 of 6) golden mice with bots compared to 14 percent (16 of 113) cotton mice with bots. Therefore, in three of these studies (Dunaway et al. 1967, Jennison et al. 2006, Pearson 1954), prevalences of larval botfly parasitism was lower in golden mice than in sympatric *Peromyscus* spp., including both studies in which golden mice and white-footed mice were compared (Dunaway et al. 1967, Jennison et al. 2006). Jennison et al. (2006) argued that golden mice are less prone to larval botfly parasitism because they have smaller, more arboreal home ranges, spend less time foraging, and occupy less varied (less three-dimensional) habitats than white-footed mice. They thought that white-footed mice would be more likely to encounter eggs of *C. fontinella* because of their more three-dimensional use of habitat space, especially at ground level compared to golden mice.

Seasonally, Dunaway et al. (1967) recorded inguinal bots in golden mice only during August. Jennison et al. (2006) recorded a unimodal peak of mature bot parasitism in sympatric golden mice and white-footed mice in mid-July (ranging from June through August) in 2001. They found bimodal peaks in mid-July and late October, respectively (ranging from June through October), in 2004 in Piedmont Georgia. They attributed the late October peak during 2004 to increased rainfall and temperatures related to an active hurricane season compared to 2001 (Jennison et al. 2006). In Coastal Plain Georgia, late-stage bots were recorded in golden mice during June and October (Durden Unpublished data), and in southern Mississippi, the single recorded bot was collected from a golden mouse in November (Clark and Durden Unpublished data). Based on these records, either one (June–August) or two (June–August followed by October–November) periods of parasitism by mature botfly larvae occur in golden mice, with a second peak more likely in more southern latitudes where winters are typically shorter and milder. Studies of *C. f. fontinella* bots parasitizing *Peromyscus* spp. support this assumption. There are more bot generations per year from north to south within the eastern United States. Univoltinism occurs in white-footed mice in Maryland (Durden 1992), bivoltinism in cotton mice in Coastal Georgia (Durden 1995), and bivoltinism or multivoltinism in cotton mice in southern Florida (Bigler and Jenkins 1975).

Vector-Borne Diseases Associated with Golden Mice

In this section, pathogens, especially zoonotic ones, that can be transmitted by ectoparasites of golden mice are considered. Particular attention is given to the two pathogens for which golden mice are known to be reservoir hosts. Botfly larvae do not fit into this category. It should be noted, however, that myiasis (infestation with fly larvae) by *Cuterebra* spp. has been documented in humans in North America, including geographical regions where golden mice occur (Baird et al. 1982, Keth 1999, Rice and Douglas 1972, Salomon et al. 1970). Nevertheless, humans are accidental hosts for *Cuterebra* spp. larvae and zoonotic cases are rare (Catts 1982).

Based on seroconversion or agent isolation from blood, only two vector-borne zoonotic pathogens are currently known to use golden mice as reservoir hosts, namely *Rickettsia rickettsii*, which causes Rocky Mountain spotted fever, and *Borrelia burgdorferi*, which causes Lyme disease (Lyme borreliosis) (Bozeman et al. 1967, Durden et al. 2004, Kollars et al. 1996). Further, evidence of these agents in golden mice has only been reported from a few regions. This compares to seven vector-borne pathogens recorded from the often sympatric white-footed mouse and five from the cotton mouse (Kollars and Durden Unpublished data). Partly for this reason, Durden et al. (2004) considered golden mice to be far less important for the enzootic maintenance of vector-borne zoonotic pathogens than are cotton mice. The same would appear to be true in this regard when golden mice are compared with white-footed mice.

Another probable reason why golden mice are less important as reservoir hosts of zoonotic vector-borne pathogens than these two species of *Peromyscus* is because golden mice are parasitized by fewer of the vectors for these pathogens. For example, in Coastal Plain Georgia, no blacklegged ticks (*I. scapularis*) and an average of only 1.9 American dog ticks (*D. variabilis*) per infested golden mouse was recorded compared to corresponding values of 1.0 and 3.0 for these two tick species on sympatric cotton mice (Durden et al. 2004). In fact, only Durden et al. (2000) in northwestern Florida have recorded *I. scapularis* (larvae only) from golden mice. Both tick species are important vectors of zoonotic pathogens to humans. Significantly, *I. scapularis* is a vector of the agents of Lyme borreliosis, human granulocytic anaplasmosis (HGA) (formerly named human granulocytic ehrlichiosis [HGE]), Q fever, tularemia, and human babesiosis. It is also a vector of less widespread agents such as deer tick virus (a strain of Powassan virus). This tick could also be a vector of other spirochetes and of one or more species of *Bartonella* (Allan 2001, Oliver 1996, Oliver et al. 2003, Sonenshine et al. 2002). *Dermacentor variabilis* is the principal vector of the rickettsial agent of Rocky Mountain spotted fever in eastern North America and can also transmit the agents of Q fever and tularemia (Allan 2001, Norment et al. 1985, Sonenshine et al. 2002). Some engorging adult females of this species can cause a reversible (if the tick is removed) ascending flaccid paralysis especially in dogs and humans (Sonenshine et al. 2002).

Immature stages of a third species of ixodid tick, *I. minor*, that has relevance to a zoonotic vector-borne disease have also been collected from golden mice in Florida and Georgia (Table 10.1). *Ixodes minor* is not known to feed on humans, but is an enzootic vector of the agent of Lyme borreliosis between rodents (and possibly between birds as well) (Oliver 1996, Oliver et al. 2003). Bridge vectors such as *I. scapularis* could feed on rodents such as golden mice (or on birds) infected by *I. minor* and then, after molting to a subsequent life stage, feed on humans to transmit the pathogen. Statistically, there was no significant difference between the infestation prevalences (3.5 percent on cotton mice, 4.4 percent on golden mice) or mean intensities (1.4 on cotton mice, 1.0 on golden mice) for *I. minor* on these sympatric rodents in Coastal Georgia (Durden et al. 2004). The golden mouse could, therefore, be as important as the cotton mouse in parts of the coastal southeastern United States regarding its role as an enzootic reservoir host for *B. burgdorferi*, although in most areas, population densities of golden mice are much less than cotton mice.

A fourth species of ixodid tick with medical relevance, *H. leporispalustris*, which parasitizes rabbits (all tick stages) and birds (immature tick stages only) (Bishopp and Trembley 1945, Cooley 1946), has been reported from golden mice (Table 10.1). This tick is an important enzootic vector of the agent of tularemia ("rabbit fever"), especially among rabbits (Allan 2001). However, this was a single tick record from the golden mouse (Bishopp and Trembley 1945), which clearly is an atypical host for this tick. It is highly unlikely that the golden mouse has an important role in the maintenance of tularemia in nature.

Another hematophagous ectoparasite with medical relevance that parasitizes golden mice is the tropical rat mite, *O. bacoti* (Table 10.1). This macronyssid mesostigmatid mite will feed on human blood if given the opportunity and can cause a characteristic rash called "tropical rat mite dermatitis" (Durden et al. 2004). *O. bacoti* is also a laboratory vector of the agents of murine typhus, rickettsialpox, Q fever, plague, and tularemia (Yunker 1973). Its importance as a vector of any of these agents in nature, however, is unknown and is probably very low or nil. Neither of the species of chiggers collected from golden mice is known to feed on humans (Farrell 1956). Considering the other mesostigmatid mites reported, *Androlaelaps* spp. might rarely feed on humans with little or no impact (Radovsky 1985).

Overall, because only two vector-borne zoonotic pathogens (from restricted geographical areas) and low numbers of the vectors for these pathogens have been reported from golden mice, this small mammal species appears to have little importance in the epidemiology of the diseases caused by these pathogens (Durden et al. 2004). However, screening of golden mice for pathogens and vectors in additional geographical areas needs to be completed before more definitive statements can be made. For example, like certain species of *Peromyscus*, golden mice could harbor other zoonotic vector-borne pathogens and parasites, such as those that cause HGA and babesiosis.

Concluding Remarks

Golden mice are parasitized or colonized by 25 species of ectoparasites or epifaunistic arthropods and by the larvae of one species of botfly. Future research on golden mice will undoubtedly add to this number, especially for parasitic and phoretic mites. The number of arthropods and their infestation prevalences and abundance on golden mice are generally much lower than for sympatric, similarly sized rodents such as cotton mice and white-footed mice. This appears to be a consequence of smaller home range size, less three-dimensional use of habitat space, lower patterns of activity, and more frequent occupation of arboreal nests by golden mice compared to sympatric mice. Only two vector-borne pathogens have been recorded from golden mice, although both cause zoonotic diseases; Rocky Mountain spotted fever and Lyme disease. In comparison, white-footed mice and cotton mice support more vector-borne zoonotic pathogens (seven and five, respectively). The comparatively low number of zoonotic vector-borne pathogens collected from golden mice, together with their low number of proven vectors (three species of ixodid ticks), strongly suggest that *O. nuttalli* is much less important than some species of *Peromyscus* in the epidemiology and maintenance of zoonotic vector-borne diseases. Future ecological, epidemiological, laboratory, and integrative studies of the interactions among golden mice, their ectoparasites, and vector-borne pathogens should provide additional insight into parasitism and disease ecology for this unique small mammal.

Acknowledgments. Aspects of the work reported here were supported by NIH grants AI 24899 and AI 40729 and by a Georgia Southern University Faculty Research Grant. Fieldwork by Craig W. Banks, Barbara L. Belbey, Kerry C. Clark, Renjie Hu, Thomas M. Kollars, Jr., Todd N. Nims, and Will K. Reeves is gratefully acknowledged. Editorial input from Gary W. Barrett and George A. Feldhamer is also appreciated.

Literature Cited

Allan, S.A. 2001. Ticks (Class Arachnida: Order Acarina). Pages 72–106 *in* W.M. Samuel, M.J. Pybus, and A.A. Kocan, editors. Parasitic diseases of wild mammals, 2nd ed. Iowa State University Press, Ames, Iowa.

Baird, C.R., J.K. Podgore, and C.W. Sabrosky. 1982. *Cuterebra* myiasis in humans: Six new case reports from the United States with a summary of known cases (Diptera: Cuterebridae). Journal of Medical Entomology 19:263–267.

Baron, R.W., and D.D. Colwell. 1991. Mammalian immune responses to myiases. Parasitology Today 7:353–355.

Benton, A.H. 1980. An atlas of the fleas of the eastern United States. Marginal Media, Fredonia, New York.

Bigler, W.J., and J.H. Jenkins. 1975. Population characteristics of *Peromyscus gossypinus* and *Sigmodon hispidus* in tropical hammocks of south Florida. Journal of Mammalogy 56:633–644.

Bishopp, F.C., and H.L. Trembley. 1945. Distribution and hosts of certain North American ticks. Journal of Parasitology 31:1–54.

Bozeman, F.M., A. Shirai, W. Humphries, and H.S. Fuller. 1967. Ecology of Rocky Mountain spotted fever 2. Natural infection of wild mammals and birds in Virginia and Maryland. American Journal of Tropical Medicine and Hygiene 16:48–59.

Catts, E.P. 1982. Biology of New World botflies: Cuterebridae. Annual Review of Entomology 27:313–338.

Christopher, C.C., and G.W. Barrett. 2006. Coexistence of white-footed mice (*Peromyscus leucopus*) and golden mice (*Ochrotomys nuttalli*) in a southeastern forest. Journal of Mammalogy 87:102–107.

Clark, K.L., and L.A. Durden. 2002. Parasitic arthropods of small mammals in Mississippi. Journal of Mammalogy 83:1039–1048.

Clark, K.L., J.H. Oliver, Jr., J.M. Grego, A.M. James, L.A. Durden, and C.W. Banks. 2001. Host associations of ticks parasitizing rodents at *Borrelia burgdorferi* enzootic sites in South Carolina. Journal of Parasitology 87:1379–1386.

Cogley, T.P. 1991. Warble development by the rodent bot *Cuterebra fontinella* (Diptera: Cuterebridae) in the deer mouse. Veterinary Parasitology 38:275–288.

Cooley, R.A. 1946. The genera *Boophilus, Rhipicephalus*, and *Haemaphysalis* (Ixodidae) of the New World. National Institue of Health Bulletin 187:1–54.

Dunaway, P.B., J.A. Payne, L.L. Lewis, and J.D. Story. 1967. Incidence and effects of *Cuterebra* in *Peromyscus*. Journal of Mammalogy 48:38–51.

Durden, L.A. 1992. Parasitic arthropods of sympatric meadow voles and white-footed mice at Fort Detrick, Maryland. Journal of Medical Entomology 29:761–766.

Durden, L.A. 1995. Botfly (*Cuterebra fontinella fontinella*) parasitism of cotton mice (*Peromyscus gossypinus*) on St. Catherines Island, Georgia. Journal of Parasitology 81:787–790.

Durden, L.A. 2002. Lice (Phthiraptera). Pages 45–65 *in* G. Mullen and L. Durden, editors. Medical and veterinary entomology. Academic Press, San Diego, California.

Durden, L.A. 2006. Taxonomy, host associations, life cycles and vectorial importance of ticks parasitizing small mammals. Pages 91–102 *in* S. Morand, B. Krasnov, and R. Poulin, editors. Micromammals and macroparasites: From evolutionary ecology to management. Springer, Tokyo, Japan.

Durden, L.A., R. Hu, J.H. Oliver, Jr., and J.E. Cilek. 2000. Rodent ectoparasites from two locations in northwestern Florida. Journal of Vector Ecology 25:222–228.

Durden, L.A., and T.M. Kollars, Jr. 1992. An annotated list of the ticks (Acari: Ixodoidea) of Tennessee, with records of four exotic species for the United States. Bulletin of the Society for Vector Ecology 17:125–131.

Durden, L.A., and T.M. Kollars, Jr. 1997. The fleas (Siphonaptera) of Tennessee. Journal of Vector Ecology 22:13–22.

Durden, L.A., T.M. Kollars, Jr., S. Patton, and R.R. Gerhardt. 1997. Sucking lice (Anoplura) of mammals of Tennessee. Journal of Vector Ecology 22:71–76.

Durden, L.A., and G.G. Musser. 1994a. The sucking lice (Insecta: Anoplura) of the world: A taxonomic checklist with records of mammalian hosts and geographical distributions. Bulletin of the American Museum of Natural History 218:1–90.

Durden, L.A., and G.G. Musser. 1994b. The mammalian hosts of the sucking lice (Anoplura) of the world: A host–parasite list. Bulletin of the Society for Vector Ecology 19:130–168.

Durden, L.A., R.N. Polur, T. Nims, C.W. Banks, and J.H. Oliver, Jr. 2004. Ectoparasites and other epifaunistic arthropods of sympatric cotton mice and golden mice: Comparisons and implications for vector-borne zoonotic diseases. Journal of Parasitology 90:1293–1297.

Durden, L.A., and R. Traub. 2002. Fleas (Siphonaptera). Pages 103–125 *in* G. Mullen and L. Durden, editors. Medical and veterinary entomology. Academic Press, San Diego, California.

Durden, L.A., W. Wills, and K.C. Clark. 1999. The fleas (Siphonaptera) of South Carolina with an assessment of their vectorial importance. Journal of Vector Ecology 24:171–181.

Ewing, H.E. 1938. North American mites of the subfamily Myobiinae, new subfamily (Arachnida). Proceedings of the Entomological Society of Washington 40:180–197.

Fain, A. and K.E. Hyland, Jr. 1985. Evolution of astigmatid mites on mammals. Pages 641–658 *in* K.C. Kim, editor. Coevolution of parasitic arthropods and mammals. John Wiley & Sons, New York, New York.

Fain, A., and J.O. Whitaker, Jr. 1973. Phoretic hypopi of North American mammals (Acarina: Sarcoptiformes, Glycyphagidae). Acarologia 15:144–170.

Farrell, C.E. 1956. Chiggers of the genus *Euschöngastia* (Acarina: Trombiculidae) in North America. Proceedings of the United States National Museum 106:85–235.

Furman, D.P. 1959. Feeding habits of symbiotic mesostigmatid mites of mammals in relation to pathogen-vector potentials. American Journal of Tropical Medicine and Hygiene 8:5–12.

Gingrich, R.E. 1979. Effects of some factors on the susceptibility of *Peromyscus leucopus* to infestation by larvae of *Cuterebra fontinella* (Diptera: Cuterebridae). Journal of Parasitology 65:288–292.

Goodpaster, W.W., and D.F. Hoffmeister. 1954. Life history of the golden mouse, *Peromyscus nuttalli*, in Kentucky. Journal of Mammalogy 35:16–27.

Hays, K.L., and F.E. Guyton. 1958. Parasitic mites (Acarina: Mesostigmata) from Alabama mammals. Journal of Economic Entomology 51:259–260.

Hu, R., L.A. Durden, and J.H. Oliver, Jr. 2000. Winter ectoparasites of mammals in the northeastern piedmont area of Georgia. Journal of Vector Ecology 25:23–27.

Hunter, D.M., and J.M. Webster, 1973. Determination of the migratory route of botfly larvae, *Cuterebra grisea* (Diptera, Cuterebridae) in deermice. International Journal for Parasitology 3:311–316.

Jennison, C.A., L.R. Rodas, and G.W. Barrett. 2006. *Cuterebra fontinella* parasitism on *Peromyscus leucopus* and *Ochrotomys nuttalli*. Southeastern Naturalist 5:157–164.

Keth, A.C. 1999. Three incidents of human myiasis by rodent *Cuterebra* (Diptera: Cuterebridae) larvae in a localized region of western Pennsylvania. Journal of Medical Entomology 36:831–832.

Kim, K.C., H.D. Pratt, and C.J. Stojanovich. 1986. The sucking lice of North America: An illustrated manual for identification. The Pennsylvania State University Press, University Park, Pennsylvania.

Kollars, T.M., Jr., L.A. Durden, and J.H. Oliver, Jr. 1997. Fleas and lice parasitizing mammals in Missouri. Journal of Vector Ecology 22:125–132.

Kollars, T.M., Jr., J.H. Oliver, Jr., E.J. Masters, P.G. Kollars, and L.A. Durden. 2000. Host utilization and seasonal occurrence of *Dermacentor* species (Acari: Ixodidae) in Missouri, USA. Experimental and Applied Acarology 24:631–643.

Kollars, T.M., Jr., D.D. Ourth, T.D. Lockey, and D. Markowski. 1996. IgG antibodies to *Borrelia burgdorferi* in rodents in Tennessee. Journal of Spirochetal and Tick-Borne Diseases 3:130–134.

Krantz, G.W. 1978. A manual of acarology, 2nd ed. Oregon State University Bookstores, Corvallis, Oregon.

Layne, J.N. 1958. Records of fleas (Siphonaptera) from Illinois mammals. Natural History Miscellanea 162:1–7.

Layne, J.N. 1963. A study of the parasites of the Florida mouse, *Peromyscus floridanus*, in relation to host and environmental factors. Tulane Studies in Zoology 11:1–27.

Layne, J.N. 1971. Fleas (Siphonaptera) of Florida. Florida Entomologist 54:35–51.

Linzey, D.W. 1968. An ecological study of the golden mouse, *Ochrotomys nuttalli*, in the Great Smoky Mountains National Park. American Midland Naturalist 79:320–345.

Linzey, D.W., and A.V. Linzey. 1968. Mammals of the Great Smoky Mountains National Park. Journal of the Elisha Mitchell Scientific Society 84:384–414.

Linzey, D.W., and R.L. Packard. 1977. *Ochrotomys nuttalli*. Mammalian Species 75:1–6.

McCarley, W.H. 1958. Ecology, behavior and population dynamics of *Peromyscus nuttalli* in eastern Texas. Texas Journal of Science 10:147–171.

Melville, H. 1988. Moby-Dick: Or the whale. Page 456 *in* The writings of Herman Melville, Vol. 6. H. Harrison, H. Parker, and G.T. Tanselle, editors. Northwestern University Press and the Newberry Library, Evanston and Chicago, Illinois. (Originally published in 1851).

Morlan, H.B. 1952. Host relationships and seasonal abundance of some southwest Georgia ectoparasites. American Midland Naturalist 48:74–93.

Morzillo, A.T., G.A. Feldhamer, and M.C. Nicholson. 2003. Home range and nest use of the golden mouse (*Ochrotomys nuttalli*) in southern Illinois. Journal of Mammalogy 84:553–560.

Nims, T.N., L.A. Durden, C.R. Chandler, and O.J. Pung. 2007. Parasitic and phoretic arthropods of the old-field mouse (*Peromyscus polionotus*) from burned habitats with additional ectoparasite records from the eastern harvest mouse (*Reithrodontomys humulis*) and southern short-tailed shrew (*Blarina carolinensis*). Comparative Parasitology (In press).

Nims, T.N., L.A. Durden, and R.L. Nims. 2004. New state and host records for the phoretic fur mite, *Glycyphagus hypudaei* (Acari: Glycyphagidae). Journal of Entomological Science 39:470–471.

Noël, S., N. Tessier, B. Angers, D.M. Wood, and F.-J. LaPointe. 2004. Molecular identification of two species of myiasis-causing *Cuterebra* by multiplex PCR and RFLP. Medical and Veterinary Entomology 18:161–166.

Norment, B.R., L.S. Stricklin, and W. Burgdorfer. 1985. Rickettsia-like organisms in ticks and antibodies to spotted fever-group rickettsiae in mammals from northern Mississippi. Journal of Wildlife Diseases 21:125–131.

Nutting, W.B. 1985. Prostigmata–Mammalia: Validation of coevolutionary phylogenies. Pages 569–640 *in* K.C. Kim, editor. Coevolution of parasitic arthropods and mammals. John Wiley & Sons, New York, New York.

Oliver, J.H., Jr. 1996. Lyme borreliosis in the southern United States: A review. Journal of Parasitology 82:926–935.

Oliver, J.H., Jr., T. Lin, L. Gao, K.L. Clark, C.W. Banks, L.A. Durden, A.M. James, and F.W. Chandler, Jr. 2003. An enzootic transmission cycle of Lyme borreliosis spirochetes in the southeastern United States. Proceedings of the National Academy of Science USA 100:11,642–11,645.

Payne, J.A., and G.E. Cosgrove. 1966. Tissue changes following *Cuterebra* infestation in rodents. American Midland Naturalist 75:205–213.

Pearson, P.G. 1954. Mammals of Gulf Hammock, Levy County, Florida. American Midland Naturalist 51:468–480.

Pfitzer, D.W. 1950. A manual of the fleas of Tennessee. MS thesis, University of Tennessee, Knoxville, Tennessee.

Radovsky, F.J. 1985. Evolution of mammalian mesostigmate mites. Pages 441–504 *in* K.C. Kim, editor. Coevolution of parasitic arthropods and mammals. John Wiley & Sons, New York, New York.

Reeves, W.K., L.A. Durden, C.M. Ritzi, K.R. Beckham, P. Super, and B.M. OConnor. 2006. Ectoparasites of vertebrates in the Great Smoky Mountains National Park, USA. Zootaxa 1392:31–68.

Rice, P.L., and G.W. Douglas. 1972. Myiasis in man by *Cuterebra* (Diptera: Cuterebridae). Annals of the Entomological Society of America 65:514–516.

Ritzi, C.M. 2002. New ectoparasite records from two rodents from North Carolina. Journal of the North Carolina Academy of Science 118:243–245.

Sabrosky, C.W. 1986. North American species of *Cuterebra*, the rabbit and rodent botflies (Diptera: Cuterebridae). Thomas Say Foundation Monographs, Vol. 11. Entomological Society of America, College Park, Maryland.

Salomon, P.F., E.P. Catts, and W.G. Knox. 1970. Human dermal myiasis caused by rabbit bot fly in Connecticut. Journal of the American Medical Association 213:1035–1036.

Sanford, L.G., and K.L. Hays. 1974. Fleas (Siphonaptera) of Alabama and their host relationships. Alabama Agricultural Experiment Station Auburn, Alabama Bulletin 458:1–42.

Shiffer, C.N. 1983. Aggregation behavior of adult *Cuterebra fontinella* (Diptera: Cuterebridae) in Pennsylvania, USA. Journal of Medical Entomology 20:365–370.

Sillman, E.I., and M.V. Smith. 1959. Experimental infestation of *Peromyscus leucopus* with larvae of *Cuterebra angustifrons*. Science 130:165–166.

Smith, D.H. 1978a. Effects of botfly (*Cuterebra*) parasitism on activity patterns of *Peromyscus maniculatus* in the laboratory. Journal of Wildlife Diseases 14:28–39.

Smith, D.H. 1978b. Vulnerability of botfly (*Cuterebra*) infected *Peromyscus maniculatus* to shorttail weasel predation in the laboratory. Journal of Wildlife Diseases 14:40–51.

Sonenshine, D.E., R.S. Lane, and W.L. Nicholson. 2002. Ticks (Ixodida). Pages 517–558 *in* G. Mullen and L. Durden, editors. Medical and veterinary entomology. Academic Press, San Diego, California.

Strandtmann, R.W. 1949. The blood-sucking mites of the genus *Haemolaelaps* (Acarina: Laelaptidae) in the United States. Journal of Parasitology 35:325–352.

Timm, R.M., and E.F. Cook. 1979. The effect of botfly larvae on reproduction in white-footed mice (*Peromyscus leucopus*). American Midland Naturalist 101:211–217.

Wecker, S.C. 1962. The effects of botfly parasitism on a local population of the white-footed mouse. Ecology 43:561–565.

Whitaker, J.O., Jr. 1968. Parasites. Pages 254–311 *in* J.A. King, editor. Biology of *Peromyscus* (Rodentia), American Society of Mammalogists, Special Publication 2. American Society of Mammalogists, Lawrence, Kansas.

Whitaker, J.O., Jr., G.S. Jones, and D.D. Pascal. 1975. Notes on mammals of the Fires Creek area, Nantahala Mountains, North Carolina, including their ectoparasites. Journal of the Elisha Mitchell Scientific Society 91:13–17.

Whitaker, J.O., Jr,, and N. Wilson. 1974. Host and distribution lists of mites (Acari), parasitic and phoretic, in the hair of wild mammals of North America, north of Mexico. American Midland Naturalist 91:1–67.

Yunker, C.E. 1973. Mites. Pages 425–492 *in* R.J. Flynn, editor. Parasites of laboratory animals. Iowa State University Press, Ames, Iowa.

Section 4
New Perspectives and Future Challenges

11
Aesthetic Landscapes of the Golden Mouse

Terry L. Barrett and Gary W. Barrett

We are faced with some serious multi-disciplinary and multi-cultural dilemmas. Some of these are methodological, as well as philosophical and political. . . How can spiritual and cultural values be incorporated into planning and policy decisions? Can any of this be assigned monetary value? If not, how can other value systems be respected and weighed? (Posey 1999:549)

Defining Ecological Aesthetics

Aesthetics, as a transcending process, is based primarily on the economy of, and relationship between, natural and cultural systems. The German zoologist Ernst Haeckel (1869) defined ecology (the study of the "house" based on the Greek *oikos*) as a body of knowledge concerning the economy of nature (i.e., the total relationships of both the organic and the inorganic environments). Traditionally, aesthetic has been defined as "a philosophical theory or idea of what is aesthetically valid at a given time and place." *Economy* is defined as (1) "the efficient, sparing, or concise use of something" and (2) "the management of the resources of a community, country, . . . especially with a view to its productivity" (*Webster's Unabridged Dictionary of the English Language*, 2nd ed., s.vv. "Aesthetics," "Economy"). In this chapter, we define *aesthetics* as an economy transcending all natural and artificial levels of organization, encompassing what human beings perceive as a resource and a nonresource in a given place and time (Barrett et al. 1999). What humans perceive as a resource and a nonresource in a given place and period of time is about *an economy of survival*—survival on political, commercial, social, and artistic grounds.

The epistemology, proposed here, precludes the notion of aestheticism, defined as (1) the acceptance of artistic beauty and taste as a fundamental standard, with ethical and other standards being secondary, and (2) an exaggerated or excessive devotion to art, music, or poetry, with indifference to practical matters (*Webster's Unabridged Dictionary of the English Language*, 2nd ed., s.v. "Aestheticism"). This epistemology places increased emphasis on the role of aesthetics as a primary process that transcends all levels of organization (Barrett et al. 1997). This revised definition suggests a framework in which aesthetics is viewed as a transcending process, not only within the traditional context—that is, concerning artistic beauty—but also as an emergent property based on the coevolution of artificial (anthropogenic) and natural (ecological) systems (Cairns 1997). In this way, aesthetics can be understood more broadly as an economy for survival, regarding such concepts as carrying capacity or self-organization. This definition allows for a more effective management of *natural capital*, meaning the benefits and services that are supplied to societies by natural ecosystems (Daily et al. 1997).

Landscape Aesthetics as a Model

Researchers attempt to quantify landscape aesthetics by integrating blocks of data (Farina 2000, Zonneveld 1995) in order to determine significant differences, based on level of probability, in search of a "universal" aesthetic. Aesthetics as a transcending process, through levels of organization, is traditionally viewed as a constant: the indicator of the societal canon of taste (Gans 1999). This view is conducive to existing quantitative methods of investigation, such as the survey (Nassauer 1988).

As Werner (1999:102) explained, "One difficulty with . . . universalist models is that their simplified, typically stationary representation of external forcing is at odds with the forcing and resulting response of natural systems." The subject of quality then, regarding landscape aesthetics, has been addressed primarily through philosophical approaches. Aesthetics becomes a transcending property of transcending processes, including behavior, development, diversity, energetics, evolution, integration, and regulation. Then aesthetics emerges as transcendental in nature (Barrett 1997). Exemplary of aesthetics, as a transcendental position, is the intuitive and spiritual land ethic espoused by Leopold (1949) and Emerson (Metzger 1954), or the material logic of Kant (Baxley 2005, Swing 1969).

Bourassa (1991) maintained landscape as object, although he suggested that the aesthetics of landscape is the experience of the interaction between object and subject. Therefore, investigations exclusively based on either quantitative or qualitative aspects of landscape are inadequate. Figure 11.1 shows vertical (levels of organization) and horizontal (specialization within levels) relationships of natural and artificial systems, providing a model in which aesthetics, as a process, transcends all levels of organization of natural systems: the cellular, organismic, and ecological (after Barrett et al. 1997). Specialization within levels is, respectively, alive (i.e., being capable of vital functions; *oblivious*), self-aware (i.e., externalizing; *conscious*), and self-reflective (i.e., mirroring; *consideration of self*). Levels of growth and development, at increasing levels of organization, relate to a dual hierarchy illustrating the dependent, independent, and interdependent levels of individual growth (Covey 1989) and of cultural development (Clarke 2000). Organisms, including the golden mouse (*Ochrotomys nuttalli*), become more interdependent at higher levels of organization.

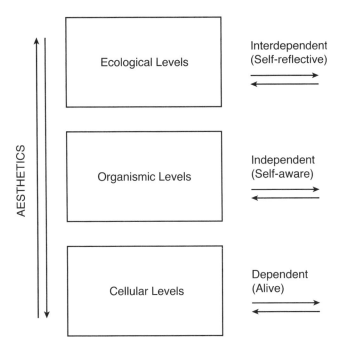

FIGURE 11.1. Aesthetics as a transcending process with concepts restricted to individual levels of organization. After Barrett et al. 1997.

Aesthetic Landscapes as Concept

> Since the period of antiquity until modern times the concept of nature has undergone changes in several ways. One of the more fundamental points, as has been discussed (e.g., by Collingwood *[The Idea of Nature: by R. G. Collingwood 1945]*), is that nature in the earliest times had its basic quality in the aspect of something being born (compare e.g., the French word *naitre*). Thus, according to this view, nature is the potential state from which another state could emerge. (Karlqvist and Svedin 1993:5)

Karlqvist and Svedin (1993:5) continue: "This view of nature emphasizes "potentiality" aspects which are distinctly different from the 'modern' views emerging in Europe during the seventeenth-century. Here, for the first time nature is referred to as a 'landscape,' (i.e., an object and a thing) . . ." The representation of landscape for its own sake is associated with the detailed paintings of the American wilderness between 1825 and 1875. Although "landscape" is defined as a picture showing natural scenery, without narrative content (Kleiner et al. 2001:1156), the descriptions of the new world, through these evocative and picturesque portraits, do not appear to escape the spiritual and moral prejudices in the agendas of this period. Conversely, landscape hierarchized as a marginalized topography (Andrews 1999) is exemplified in the diptych *Portrait of Federico da Montefeltro and his spouse Battista Sforza* (Figure 11.2, see color plate section), painted by Piero della Francesca (circa 1420–1492) during the humanist movement (i.e., each individual had significance within society). This work reveals the naturalistic figures of Federico da Montefeltro and Gemahlin Battista Sforza situated in the foreground of a realistic landscape. This work, circa 1465, views the influence of individual nobility upon a backdrop of landscape.

Nietzsche would interpret each of these aesthetic landscapes as a "falsification" of the world because

> . . . the world we are acquainted with through sensation and perception is a unique world of qualities, one that is "true" for us as a transmuted humanized world of experience . . . The *naiveté* of some thinkers, he remarks, was their failure to see that our senses and our "categories of reason" involved "the adjustment of the world for utilitarian ends." In effect, an "anthropocentric idiosyncrasy" was taken as the measure of all things. (Stack 2005:9)

In similar context, Ahl and Allen (1996) extrapolate that

> . . . observation differentiates between figure and ground. Figure is that part of the observation field that is treated as significant, and ground is everything else. According to this model, the boundary between figure and ground is not given by the external world. Both figure and ground is the product of an observer's assertions, questions, values, prior beliefs, and expectations. (Ahl and Allen 1996:37)

Consequently, species within these landscapes, such as the golden mouse, regarded as neither the deviant nor exotic figure, have been discounted as a nonessential that lends a kind of invisibility to their value. In other words, to Ahl and Allen (1996) the golden mouse would be too small for the human gaze and becomes undifferentiated background. However, the golden mouse, as part of this undifferentiated background, is important even though initially not seen as figure.

Landscape Aesthetics: A Multidimensional Mosaic

Landscape aesthetics is a cultural decision that informs the natural environment on a global scale—influencing such diverse ecological footprints as terraces of green tea or skyscapes of red sunsets, and, yes, riparian edges of thick forest canopies nested with communities of the golden mouse. Beyond the realm of traditional thought and formal space, even *only two humans* standing in "promise" or in "secret" might change the meaning and pattern of the landscape. Turner et al. (1995) noted that there is a critical need for a more complete understanding of human driving forces upon ecological change and that such an understanding requires an interdisciplinary, problem-solving approach—an integrative approach to landscape aesthetics.

Artificial (anthropogenic) systems have developed a persona, other than *Other*, that has led to interdisciplinary fields of study such as human engineering (ergonomics), ecosystem health, and artificial intelligence. *Artificial is defined here as a logical consequence of the manipulation of the natural world by humans.* Artificial is an extension of self, evolving from the first object an individual selected to mark (i.e., projection) or capture (i.e., possession) his or her environment (Huyghe 1962).

> The instinctual capacity of human beings to use tools interacts with their intellectual capacities to make possible the external storage of information. Developments within human cultures in historical times, beginning with the invention of writing and ending currently with the development of computers, have increased the amount of information stored and available to humans beings by many orders of magnitude. (Jackson 2000:23)

This capacity to comprehend large external caches of information has expanded sophistication of *mark-projection* (*sensu* Huyghe 1962) in the forms of animatism, anthropormorphization, and personification and *capture-possession* in the forms of curio, museum, and theme park.

Aesthetic Landscape: Mark-Projection as Literature, Art, and Icon

Art, dance, literature, and music are evolved forms of mark-projection (Lock and Peters 1996) that influence the environment by artificial means. The result of this interaction is an aesthetic that has particular social benefits to each culture. *Animation*—the attribution of consciousness to inanimate objects and natural phenomena, *anthropormorphization*—ascription of human form or attributes to an animal, a plant, or a material object, and *personification*—the attribution of a personal nature or character to inanimate objects or abstract notions, especially as a rhetorical figure—are used to mark or project ourselves upon the environment.

The Golden Mouse in Character

> It is the mark of human beings that they integrate complex cultural ideas into fantasies or plans for action, and communicate these by means of language; this ability is vital to human survival, which depends on complex social actions. It is this general ability to fantasize that underlies the capacity to produce those complex and sophisticated fantasies that we call works of literature. (Jackson 2000:24)

Formal teaching of survival skills for the new world through literature began in the United States with the Normal schools. In 1894, William Holmes McGuffey included the word *RAT* with an accompanying etching of the depiction of a rat in the "picture lessons" of *McGuffey's Pictorial Primer* (see Minnich 1936:37 for illustration). By the publication of *McGuffey's Eclectic Primer*, the image of a rat was placed in context of a cat shown pouncing feline fashion, with the rat bound tightly within its grasp (see Vail 1909:8 for illustration). *McGuffey's Third*

Eclectic Reader introduced *Three Little Mice* by Julia C. R. Dorr, Lesson XXVI (Vail 1920a:67–68) that personified these small mammals as follows:

> I will tell you a story of three little mice,
> If you will keep still and listen to me,
> Who live in a cage that is cozy and nice,
> And are just as cunning as cunning can be.
>
> They look very wise, with their pretty red eyes,
> That seem just exactly like little round beads;
> They are white as the snow, and stand up in a row
> Whenever we do not attend to their needs; –
> Stand up in a row in a comical way, –
> Now folding their forepaws as if saying, "please;"
> Now rattling the lattice, as much as to say,
> "We shall not stay here without more bread and cheese."
>
> They are not at all shy, as you'll find, if you try
> To make them run up to their chamber to bed;
> If they don't want to go, why, they won't go—ah! no,
> Though you tap with your finger each queer little head . . .

The familiarity with and curiosity about these species, which is woven into contemporary literature, seems to indicate a continuing evolution in human involvement with Rodentia. *Elmo Doolan and the Search for the Golden Mouse* by Shirley Rousseau Murphy was published in 1970. The young reader is taken through the scholarly process of researching the golden mouse, culminating in a published book by a family of mice. The juxtaposition of the "family of mice" writing a book about the "golden mouse" allegorizes a sophisticated methodology of instruction for academic survival (Murphy 1970).

Anthropormorphization and personification of small mammals, such as the golden mouse, with lifestyle cues, such as the ornamentation of the female and male body; social interactions, such as the public expression of affection and celebration; and language, such as regional dialect and colloquialisms, each reflect a level of behavioral tolerance and a preference of resources that address a particular landscape aesthetic. A Harper Trophy, revised edition of the 1995 publication of *Poppy* (AVI 2007a) included in the anthropormorphic characters, Ragweed, a golden mouse (*O. nuttalli*) having dialogue with Poppy, a deer mouse (*Peromyscus maniculatus*):

> One of the two, a deer mouse, crouched cautiously beneath a length of rotten bark, the other, a golden mouse, stood in the open on his hind legs, his short tail sticking straight out behind for balance. From his left ear an earring dangled. In his paws he held a hazelnut . . . On Bannock Hill, the golden mouse turned to his timid companion and said, "Poppy, girl, this hazelnut is bad-to-the-bone. Bet you seed to sap there's more where it came from. Come on out and dig." (AVI 2007a:3)

In 1999, *Ragweed* was published as an adventure of a country mouse visiting the city. The transposed golden mouse, Ragweed, traveling from a rural to an urban environment, mimics fabled swings of the social pendulum attributed to human settlement. Complete was a gift for Ragweed, who had joined the city mice including Blinker and Clutch, in a defeat of the urban cats Silversides and Graybar, before departing the city by train:

> She reached up and removed her purple earring. She held it out so that the bead dangled from her paws. "We'd sort of like to give this to you, dude. I mean, if you want it, that is."
> Ragweed took the earring gently. He was deeply moved.
> "When you wear it, Ragweed, think of us and dance," Blinker suggested.
> "Like, long as you wear it," Clutch added. "you'll never back down to any bully."
> "I hear you," Ragweed said.
> "Want me to put it on, Dude?" Clutch asked.
> "Be way cool."
> Clutch fixed the earring to Ragweed's left ear. "Glad you came dude." She gave the same ear a nuzzle as she added, "Dude, you totally buttered the muffin."
> (AVI 2007b:203)

Revealed in this selection of literature, the Rodentia becomes more complex over periods of time with the detailed layering of multidimensional culture. In the McGuffey literature, for example, there is a cultural perception of Rodentia as a pest or deviant that is controlled by the predation of the dutiful cat, as well as Rodentia as the domesticated white mice caged in a laboratory-like environment with various behavioral responses to regiment. In the AVI book, the perception of Rodentia is the golden mouse, traversing the rural and urban landscapes in conversation with a diversity of species, illustrating relationships and differences in small mammal niches within the landscape mosaic. The following exemplary paragraphs cue differences in natural and artificial systems respectively:

> Ragweed, a golden mouse with dark orange fur, round ears, and a not very long tail, was saying goodbye to his mother and father as well as to fifty of his brothers and sisters. They were all gathered by the family nest, which was situated just above the banks of the Brook. (AVI 2007b:1)

> Once past the old rusty water pump, Poppy had to cross Old Orchard. Mr. Ocax's (owl) permission was not required here. Even better, the grass was high among the old twisted apple trees, providing good camouflage. Here and there delicate pink lady's slippers bloomed. Berry bushes were heavy with fruit. Bluebirds, jays, and warblers flitted by. Grasshoppers leaped about joyfully. (AVI 2007a:72)

Concepts, such as predation and the food web, are woven into the dialogue of the characters, such as the higher-trophic-level owl, "Mr. Ocax has been here longer than any mouse's living memory." (AVI 2007a:10)

> "Mr. Ocax protects us from creatures that eat us," Lungwort answered gravely. "Raccoons, foxes, skunks, weasels, stoats." One by one he displayed the pictures of the animals. "Most important, he protects us from *porcupines*." (AVI 2007a:11)

Human beings mark or project upon the environment by naming or describing that which is a resource or a nonresource. This forms a dynamic aesthetic by which individuals navigate or negotiate their environment. An oral or written story passes on vital information to future generations. Knowledge of a species, including behavior and niche, as presented in this literature by AVI, influences the investigation of a species through empathy and familiarity and reflects a cultural predetermination of the golden mouse as a rarity (curio).

In the southeastern United States, the landscape aesthetics has determined the golden mouse not as an animal for direct human food consumption (resource) (see Chapter 7 of this volume), and not as an animal of particular concern to human health (nonresource) as compared with similar mammalian species (see Chapter 10 of this volume), but as an intellectual curiosity. The golden mouse wearing an identifying ear tag (Figure 11.3A, see color plate section) is shown as part of a scientific investigation. Poppy, the deer mouse, is pictured donning Ragweed's earring (Figure 11.4B, see color plate section).

The Orange Colored Mouse as Image: John James Audubon

In *The Smithsonian Book of North American Mammals*, Pagels describes physical dimensions of the golden mouse as "head and body length 140–190 (165) mm, tail length 67–97 (80) mm, and weight (18–27) g" (Wilson and Ruff 1999:584). The individual golden mouse possesses an individual physical form based on the golden proportion of anatomical structure of animals (Doczi 1994). This economy of evolution marks the unique physical and social aesthetic of the golden mouse, yet shares a natural history with other species living in the undifferentiated background.

In 1846, J. T. Bowen in Philadelphia printed the work of John James Audubon in a three-volume folio entitled *Viviparous Quadrupeds of North America*. Although *Arvicola nuttallii* (*Audubon's Wildlife* 1989:130) and *Mus* (*Calomys*) *aureolus* (*Audubon's Wildlife* 1989:280) were yet to be classified as *O. nuttalli*, the naturalistic image of the golden mouse was pressed to paper, likely by John J. Audubon or his son John Woodhouse Audubon. They imposed a combination of watercolor, pastels, pencil, and oils upon the backgrounds likely painted by his son Victor Gifford Audubon. Then, from the transference of the original drawings to stone by William E. Hitchcock and R. Trombley and the impression of the prints from the plates by Mr. Bisbaugh, the prints entitled "White Footed Mouse" (see Figure 11.5 [see color plate section] for plate XL, no. 8, p. 131) and "Orange Colored Mouse" (see Figure 11.6 for plate XCV, no. 19, p. 285) were hand colored and ready for binding (*Audubon's Wildlife* 1989:7–13). Is the golden mouse, noticed by Mr. Nuttall (Harlan 1832:447), one of the small mammals, reproduced as a curiosity—a curio worthy of art in the fashion of early nineteenth-century sensibility?

Does the print "White Footed Mouse" (Figure 11.5, see color plate section) visually portray *O. nuttalli* as the figure positioned, upon a log, uppermost in a "scenic landscape," along with the companion white-footed mice (*Peromyscus*

FIGURE 11.6 Lithography entitled "Orange-Colored Mouse," 1846, by John Woodhouse Audubon, from the work of John James Audubon. Reproduced from Audubon 1989 with permission from Borders, Inc.

leucopus)? The written description that accompanies the print from *Audubon's Wildlife* (1989:130–132), including a description of nests and patterns of behavior, suggests that of the golden mouse:

> There are several nests now lying before us that were found near Fort Lee, New Jersey. They are seven inches in length and four in breath, the circumference measuring thirteen inches; they are of an oval shape and are outwardly composed of dried moss and a few slips of the inner bark of some wild grape-vine; other nests are more rounded, and are composed of dried leaves and moss. We have sometimes thought that two pair of these mice might occupy the same nest, as we possess one, nine inches in length and eight inches in diameter, which has two entrances, six inches apart, so that in such a case the little tenants need not have interfered with each other. [Note: This large nest was likely a shelter/communal nest of the golden mouse described in Chapter 9 of this volume.]

> When we first discovered this kind of nest we were at a loss to decide whether it belonged to a bird or a quadruped; on touching the bush, however, we saw the little tenant of this airy domicile escape. At our next visit she left the nest too clumsily, and made her way along the ground so slowly, that we took her up in our hand, when we discovered that she had four young about a fourth grown, adhering so firmly to the teats that she dragged them along in the manner of the jumping mouse, *Meriones Americanus*, . . . We preserved this little family alive for eighteen months, during which time the female produced several broods of young. During the day they usually concealed themselves in their nests, but as soon as it was dark they

became very active and playful, running up and down the wires of their cages, robbing each other's little store-houses of various grains that had been carried to them, and occasionally emitting the only sound we ever heard them utter—a low squeak resembling that of the common mouse. We have been informed by William Cooper, Esq., of Weehawken, New Jersey, an intelligent and close observer . . . that this species when running off with its young to a place of safety, presses its tail closely under its abdomen to assist in holding them on to the teats—a remarkable instance of the love of offspring. [Note: Geographic mappings of golden mice, see Hall 1981:722, Linzey and Packard 1977:2, Osgood 1909:224, Packard 1969:374, or Chapter 2 of this volume, p. 22 do not include Fort Lee, New Jersey. These observations in 1846 suggest that the geographic range of *O. nuttalli* included the western edge of the Hudson River in New Jersey during the seventeenth and eighteenth centuries.]

In 1846, one might have appreciated that the taxonomy and ecology of small mammals were in earlier stages of investigation. For example, the white-footed mouse (*P. leucopus*) was described and rendered as *Mus leucopus* Rafinesque at that time. It is also significant that the golden mouse was observed at that time, but possibly thought by some to be the white-footed mouse. Audubon noted that "this small mammal's favorite resorts are isolated cedars growing on the margins of damp places, where greenbriers, *Smilax rotundifolia*, and *S. herficea* connect the branches with the ground, and along the stems of which they climb expertly" (*Audubon's Wildlife* 1989:130–132). It has now been documented in numerous studies that golden mice prefer undergrowth containing *Smilax* sp. and similar type vines (Barbour 1942, Goodpaster and Hoffmeister 1954, Linzey and Packard 1977). There is little doubt that the nest described by Audubon in 1846 was that of *O. nuttalli*. " . . . We have also occasionally found their nests on bushes, from five to fifteen feet from the ground. They are in these cases constructed with nearly as much art and ingenuity as the nests of the Baltimore Oriole" (*Audubon's Wildlife* 1989:130–132). Bull and Farrano (1977) in *The Audubon Society Field Guide to North American Birds: Eastern Region* described the nest of the Baltimore Oriole (*Icerus galbula*) as a well-woven pendant bag of plant fibers, bark, and string suspended from the tip of a branch. (see Chapter 9 of this volume for a summary of similar materials used in construction of the globular nest of the golden mouse.) Figure 11.7 shows an arboreal golden mouse nest suspended in Spanish moss on Amelia Island, Florida; note the similarity in the description of this golden mouse nest and the Baltimore Oriole nest described above. The description of the collected nests in 1846 (i.e., ". . . oval shape and are outwardly composed of dried moss and a few slips of the inner bark of some wild grape-vine; other nests are more rounded, and are composed of dried leaves and moss" [*Audubon's Wildlife* 1989:130–132]) allows little speculation that these were golden mouse nests.

Accompanying "Orange Colored Mouse" (Figure 11.6) is a brief description of *Mus* (*Calomys*) *aureolus*—from Audubon and Bachman (1841:99), which was eventually referred to as *O. nuttalli* (see Chapter 2 of this volume for early

FIGURE 11.7. Arboreal nest of *O. nuttalli*, Fort Clinch State Park, Amelia Island, Florida. Photograph by James N. Layne.

taxonomy). These authors noted that this species resembles the white-footed mouse and remarked that ". . . this is the prettiest species of *Mus* inhabiting our country." Audubon and Bachman (1841) continued,

> . . . it is at the same time a great climber. We have only observed it in a state of nature in three instances in the oak forests of South Carolina. It ran up the tall trees with great agility, and on one occasion concealed itself in a hole (which apparently contained its nest), at least thirty feet from the ground. The specimen we have described was shot from the extreme branches of an oak in the dusk of the evening where it was busily engaged among the acorns. (*Audubon's Wildlife* 1989:280)

The "Orange Colored Mouse," referenced with the early patterns of behavior, color of pelage, and physical description, as proposed by the Zoological Society of London (17 February 1837), appears to be one of the first (if not the first) record-ed collected specimens of *O. nuttalli*.

In "White Footed Mouse" and "Orange Colored Mouse," individual small mammal figures are placed in a triangle with little apparent interaction or function other than a taxonomic comparison of color and anatomy. The composition elicits little participation from the viewer other than in an appeal for the appreciation, uttered by the current canon of craft and beauty at that time. In other words, the mice were appreciated within a level of tolerance for the natural landscape, which was inclusive of these small mammals. These curious figures stepped outside the undifferentiated background long enough to contribute to the nineteenth-century perception of beauty in nature.

Mickey Mouse as Icon

In 1928, Walt Disney and Ub Iwerks worked in secret on a new series about a mouse. The series was based on a curiosity—a mouse that had inhabited the Kansas City office of Walt Disney:

> It used to crawl across my desk and I'd feed it bits of cheese," he said. "I got quite fond of it and looked forward to its visits. It was so tame—and cheeky, too . . ." Later Disney would find when he had run into bad times in Kansas City, he would tell Ub and Roy, there came a period when he couldn't even afford cheese for the mouse any longer. And that was when I got worried he told them. "I was afraid I'd given it a taste for cheddar and it would get itself killed trying to steal the cheese the guys downstairs were using to bait their traps in their restaurant kitchen. So one day, I took the mouse to the woods outside of town and let it go. (Mosley 1990:102)

Walt Disney named the mouse Mortimer, only later to be renamed Mickey. Other human characteristics were ascribed to Mickey:

> Ub agreed with Walt that Mickey should be a brash, imprudent little chap; kind to old ladies, but mischievous with those who could look after themselves; and up to all sorts of tricks. Painstakingly, they built up a personality for him . . . Mickey's main trouble was he hated bullies, and his gallantry and courage were bigger than his muscles. He was always running into difficulty trying to rescue weaker people from oversized villains and getting creamed for his temerity. But it never stopped him, and his resourcefulness always won out in the end. (Mosley 1990:102)

During the Depression years of 1929 through 1940, the animated personification of a mouse named Mickey, who expounded courage and resourcefulness, played to the troubled mood of the times by reassuring that the values of a generation remained vital in surviving the despairing situation at hand.

Drawings of Mickey are shown in Figure 11.8. The anatomy of this mouse is structured on a vertical axis (erect), camouflaged with clothing appropriate to the gesture, provided with tools such as the boxing gloves and shotgun, and given expressions— a projection upon the mouse of a human response to a cultural situation. "Walt admitted that he had occasionally used his own face in the mirror as a model and that a lot of the expressions were his, though he swore the character was based on an actual mouse that had once prowled his tiny office in Kansas City" (Mosley 1990:101).

Expression is apparent to the human gaze as cues to the emotional state of the animal (Darwin 1872) however informed the intent of the human perception, however subtle the empathy (see Figure 11.9, photograph depicting golden mouse expression). This familiarity with animals has led to debate regarding the human perception that animals possess distinctive behavior that resembles humans (Barber 1994, Marzluff and Angell 2005). There is a growing collection of animal-personality studies and an Animal Personality Institute founded by Sam Gosling of the University of Texas (Siebert 2006).

Empathy with Mickey Mouse is symbolized by the donning of the Mouseketeer ears, celebrating the Golden Anniversary Mouse (Figure 11.10). The child, by wearing the Mickey Mouse ears, expounds the cultural and

Figure 11.8. Anthropormorphized and personified Mickey Mouse. Line-art illustrations detail development in the animation of the figure; note the button eyes and hosepipe limbs, probably by Ub Iwerks, evolved to include the rounded body form and moving eyes drawn by Fred Moore, 1930s (Mosley 1990). © Disney Enterprises, Inc.

individual acceptance of the character traits that are superimposed on Mickey by that same culture. This is repeated throughout history by a type of cultural shorthand in the "icon" of the mouse.

An iconic summation of American culture, perhaps symbolized as a figure resembling that of Mickey Mouse, is represented in Figure 11.11. Situated by the soldier, the graffito is layered with a spray-painted phallic symbol, and an X is placed across the face of the figure. The nonresource (pest), namely the *real mouse*, likely noticed by Walt Disney as a curiosity, emerges as a contribution to the economy of survival, not only by informing the natural environment of

FIGURE 11.9. Expressive golden mouse, HSB Experimental Site, Athens, Georgia. Photograph by Thomas Luhring.

FIGURE 11.10. Child with Mouseketeer ears, as Golden Anniversary Mouse expounds values of Walt Disney. © Disney Enterprises, Inc.

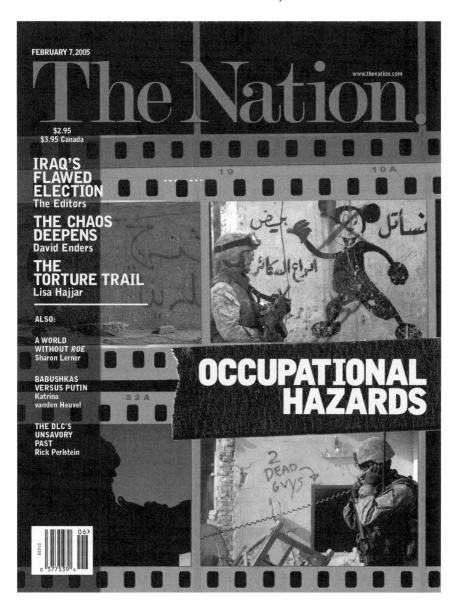

FIGURE 11.11. Front cover of *The Nation*, 7 February 2005; note a figure resembling that of Mickey Mouse, an icon of American culture, as graffito. Front cover, courtesy of The Nation Company.

biodiversity, but also as a *virtual mouse*, by compressing patterns of human behavior into congregating vignettes of cultural succession.

Each of these examples represents an escalating degree of involvement between natural and artificial systems perpetuated by distinct human projection that marks the natural environment. The stylized view of nature framed by the Audubon print, the comedic archtype in the animation of Mickey, the implicit "moral code" in the Disney crown of golden mouse ears, and the spontaneous graffiti from a subconscious socialscape, emerge as landscape aesthetics.

Aesthetic Landscape: Capture-Possession as Repositories of Information

Human beings capture or possess, by altering or collecting, that which is selected as a resource from their environment. Curiosity has facilitated the development of curios, museums, and theme parks. Repositories of information cache the living and nonliving (Ritvo 1987), and the organic and the inorganic (Lippard 1999). Archival libraries, tissue banks, preservation of linage fauna and heritage flora, and museums, are cultural repositories. The information to be stored is distilled from a cultural perception of what is valuable. The repository is a compressed vignette of these present and past valuables. For example, the golden mouse can be traced through fossil records, written observation, captivity, specimen, taxidermy, tissue sample, DNA sequence, and iconic image in any number of media. This diverse aesthetic landscape allows a single species, such as the golden mouse, to survive in multiple forms.

Aesthetic Set Points

The aesthetic set point of an individual is influenced by the cultural and historical boundaries created along with his or her life experiences. Culture and history significantly influence what human beings perceive as a resource and a nonresource (Barrett et al. 1999). This relationship represents a positive–negative feedback mechanism (Odum et al. 1995). Feedback is necessary to restore the level of tolerance or operating range of a specific society. A positive feedback is typically expressed as a resource, whereas a negative feedback might be interpreted as a nonresource.

A set-point model applicable to individual and societal aesthetics is depicted in Figure 11.12. The set point in natural ecological systems is frequently referred to as optimum carrying capacity (Barrett and Odum 2000). *An aesthetic set point within an artificial system is defined as the measure of optimum tolerance exhibited by an individual based on his/her behavior within a specific regional landscape and system(s) of belief.* Optimal behavior is defined as the best behavior that an individual can perform in the given circumstances in accordance with particular optimality criteria (McFarland 2006). In this case, the *optimality criteria* (i.e., the criteria in relation to which it is possible to determine which of a set of alternatives is best, and measured by the fitness or profitability of the species [McFarland 2006]) are being provided by the culture of that society. "The culture of a society is a kind of collective mind, or perhaps the collective matrix of any

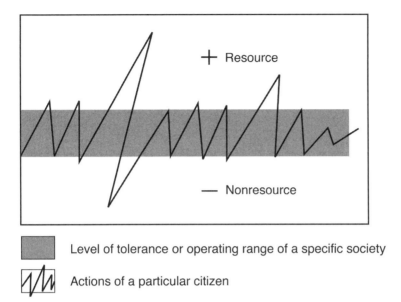

FIGURE 11.12. Model of interaction between the citizen and a specific society that determines the aesthetic set point or level of tolerance.

individual mind, and it is also a perpetual partner in dialogue for any individual mind" (Jackson 2000:23).

The aesthetic set point of a system evolves from the continuous interaction among natural and multicultural processes, transcending the individual aesthetic set point. Reciprocally, the integrity of the individual aesthetic set point might also diversify and be modified by negotiating through this multitextual and multidimensional environment. Policy and management practices are reflections of this process at the landscape scale. Biodiversity is defined as biological or physical factors that foster *adaptations*, rather than promote similarity. Artificial or cultural diversity is defined as aesthetic factors that foster *adoptions*, rather than promote similarity. *Mature systems tend to elicit more diversity of opinion and behavior within individuals and communities and, therefore, create emergent landscape properties that are dynamic and complex both ecologically and aesthetically.* "A complex society that has collapsed is suddenly, smaller, simpler, less stratified and less socially differentiated. Collapse then is not a fall to some primordial chaos, but a return to the (normal) human condition of lower complexity" (Tainter 1988:197).

The perception of an animal as *charismatic* is determined by the unique aesthetic set point of a culture, involving particular perceptions of attraction that incorporate attributes such as symmetry of facial features and accompanying behavioral traits such as cleverness or sociability. The coyote (*Canis latrans*) is perceived as charismatic in one culture or the dolphin (Family Delphinidae) in another. For example, Jungian biology theorizes that ". . . at the highest levels animals like lions and dogs do not merely behave, they act intelligently; they spend time thinking what to do; a stimulus leads directly not to action but to fantasy—imagined action—which in turn may lead to real action" (Jackson 2000:101). The human perception of the behavior of these animals creates an empathy with these animals as a "human-like" value (Ritvo 1987). Through time, the golden mouse has increasingly become accepted as an example of charismatic mammalian microfauna. This is perhaps due to the unique characteristics of sociality within and among similar species and the docile demeanor associated with the golden mouse (Christopher and Barrett 2007).

Restorative Repositories to Scale

In recent years, much funding and human effort have contributed to the restoration of large charismatic mammals (Maehr 2001, Maehr et al. 2001, Noss 2001). For example, a large-carnivore corridor has been established in Baniff National Park to increase the abundance of wolves (*Canis lupus*) and grizzly bears (*Ursus arctos*) in the Cascade Corridor (Duke et al. 2001). To date, most large-scale landscape corridors have been established to promote movement between and among landscape patches for large mammals (megafauna) such as the Florida panther (*Felis concolor coryi*) to prevent inbreeding (Perry and Perry 1994). Rosenberg et al. (1997) described the form, function, and efficacy of biological and landscape corridors with emphasis on small mammals (microfauna), such as the meadow vole (*Microtus pennsylvanicus*), the root vole (*Microtus oeconomus*), and the Eastern chipmunk (*Tamias striatus*) (Andreassen et al. 1996, Henderson et al. 1985, LaPolla and Barrett 1993). Because riparian forest corridors have been established mainly for movement of megafauna in the Florida Greenways Initiative (Harris and Scheck 1991) and the golden mouse resides in riparian habitats (Christopher and Barrett 2006, Miller et al. 2004), this corridor network initiative will also likely protect and perhaps increase the movement and abundance of golden mice and other microfauna residing in these habitats.

Hierarchy Theory: Temporal/Spatial Scaling

Hierarchy is defined here as arrangement into a graded series, such as levels of organization. A hierarchy ranging from the microcosm (Margulis and Sagan 1997, Taub 1997), mesocosm (Barrett et al. 1995, Odum 1984) to the macrocosm (natural systems) levels of ecological scale are depicted in Figure 11.13 (see color plate section). A microcosm (from Greek *mikos*, small) is a small, simplified ecosystem, such as a terrarium or aquarium. A mesocosm (from Greek *meso*, middle) is a midsized experimental ecosystem, such as large experimental fish tanks or small mammal enclosures. Macrocosm (from Greek *macros*, large) is a large natural ecosystem, such as an island or a landscape peninsula.

Kolasa and Waltho (1998) provided a detailed discussion regarding a hierarchical view of habitat and its relationship to species abundance. For example, they illustrated a three-dimensional conceptual model of habitat structure ranging from habitat specialists to habitat generalists (see Kolasa and Waltho 1998:67). The golden mouse is considered a habitat specialist compared to the white-footed mouse, which is considered a habitat generalist (Christopher and Barrett 2006, Dueser and Shugart 1978). Interestingly, specialization is correlated with the degree of habitat fragmentation in their model. The importance of landscape fragmentation and disturbance to survivorship and reproductive success of the golden mouse are discussed in Chapter 6 of this volume.

The golden mouse functions within each of these ecosystem scales. For example, Figure 11.13A (see color plate section) shows a terrarium especially designed for golden mice at HorseShoe Bend (HSB) Ecological Research Center, University of Georgia, located near Athens, Georgia. Undergraduate students studying ecology as part of field courses, teachers participating in educational workshops, and individuals interested in understanding the role of small mammals in nature use such microcosms as a learning tool—a tool necessary in the conservation of rare or unique small mammal species.

Just as zoological parks and wild natural exhibits frequently focus on large mammals, such as bears and giraffes (*Giraffa camelopardalis*), and predaceous birds, such as golden eagles (*Aquila chrysaetos*), it is significant to note that these facilities have increasingly exhibited small mammals such as bats, deer mice, and flying squirrels. The most recent exhibit of mammals in the National Museum of Natural History at the Smithsonian Institution in Washington, DC is exemplary. Importantly, the Brookfield Zoological Park located in Chicago, Illinois has an exhibit of live golden mice.

The mesocosm scale is represented in Figure 11.13B (see color plate section), which depicts a breeding area crafted for golden mice. The Brookfield Zoological Park has maintained a breeding population of golden mice in "The Swamp" area of the zoological park for over a decade. Records have been maintained on natality (to maintain the population, rather than to maximize abundance), longevity, and sex ratios. This exhibit is accessible to school children, educators, and interested persons. Andrea Zlabis, lead keeper of The Swamp, is concerned that this exhibit might be dismantled due to the nocturnal nature of the golden mouse, which precludes viewing during exhibition hours. Participants need to understand and appreciate that numerous small mammal species spend much of their time in tree cavities, in logs or underground, or in specially constructed nests (see Chapter 9 of this volume) for their protection from predators and for reproductive success. Mesocosm-scale exhibits of the golden mouse and numerous other species of small mammals inform by influencing consensus toward an aesthetic of conservation.

A 35-acre (14.1-ha) riparian peninsula created by a meander of the North Oconee River located at HSB (Figure 11.13C, see color plate section) provides an example of a macrocosm ecological system (Klee et al. 2004). Populations of golden mice and white-footed mice, among other small mammal species, inhabit this peninsula. Several studies have been conducted at this site to investigate the coexistence of two species of small mammals of similar life histories and body mass (Christopher and Barrett 2006, Jennison et al. 2006, Pruett et al. 2002). These macroscale studies have been complemented by both mesocosm-level dietary feeding studies using experimental tanks and microcosm-level bioenergetic investigations using metabolism units. Exhibits and investigations across scales (microcosm, mesocosm, and macrocosm), as well as across levels of organization (see Figure 1.4), assist in providing the knowledge and understanding to protect and conserve golden mice, as well as other species inhabiting natural ecological systems and landscapes.

The Golden Mouse as Mosaic

From antiquity, before the European "modern-time" view of landscape as object, there has returned, in the twenty-first century, a time when one might view the chromosomal structure of a living being with every hope of its "being born." Nature is rediscovered as the realm "of potential state" from which another state could emerge (Karlqvist and Svedin 1993). Patton and Hsu (1967) reported on the karyotype of a male *Peromyscus (Ochrotomys) nuttalli*. This is the golden mouse sifted through technology into a fine grain of pattern. Genetic sequencing is an aesthetic that presently reflects and informs on a global scale. In the current economy of survival, a species, such as the golden mouse, might be suspended in the compressed order of DNA (Crevar 2000).

In 1937, E. J. Moughton discovered the Reddick fossil beds that are located in Reddick I quarry, 1 mile (1.6 km), southeast of Reddick, Marion County, Florida. This site exposed the Pleistocene "remains of small rodents which Gut and Ray (1963) tentatively identified as *Peromyscus polionotus*, *P. floridanus*, *P. gossypinus*, and *Ochrotomys nuttalli* [and] are abundantly represented in the Reddick [fossil] fauna" (Pinkham 1971:28). In this "natural stele of fauna and flora" is recorded the traces of the golden mouse.

Encoded in the aesthetics of a postmodern landscape, the golden mouse is uniquely perceived as a living component of biodiversity, and a natural print of potential rebirth. Contemporary human beings stand "in promise" with their cache in repositories, as, once, Egyptians standing "in secret," entombed the meticulously mummified bodies of their pets and sacrificial animals. "It is important to recognize that . . . extinction is a multi-stage process" (Simberloff 1994:168). At present, the real golden mouse lives in the golden mouse utopia of the natural canopy, oblivious to the cyberspace of amber surrounding it.

Landscape Aesthetics: Management of Natural Capital

There has been increased interest in natural capital and in quantifying anthropomorphic influences upon natural systems. (e.g., those systems in which golden mice survive and reproduce). Quantifying the relationship of natural capital to human services (Costanza et al. 1997, Daily et al. 1997, Merkel 1998) has been human fare for centuries, from the grander of *The Silk Road* (Morris 1967) to the electronically serviced, globally linked cities of *e-topia* (Mitchell 2000). Considering the field of aesthetics as an economy of survival might allow humans to *"externalize the internalities, that is to put the contributions of economy on the same basis as the work of the environment"* (Odum and Odum 2000:21), therefore providing an equation that is simpatico with evaluation of cultural and natural capital. Aesthetics of cultural (intragenerational) and historical (intergenerational) succession evolves toward mature societies much the same as sustainable natural ecological systems evolve and develop (Barrett 1989, Lubchenco et al. 1991, Odum 1969, Odum and Barrett 2005).

The health of the individual and the ecological system continues to be influenced by and, in turn, influences aesthetics. *What people perceive as beautiful is the utopian situation of their particular culture based on the health of their natural circumstance or environment, and is repeated through their history.* What enhances the health of the ecological system increases the efficiency (i.e., the survivorship and reproductive success of the individual) (Daily et al. 2000). Economy (energetic efficiency) at all levels of organization is appealing to humans. Each perception is an aesthetic archetype, designed and manufactured by a consensus, to equip the individual with survival skills for a particular situation and present circumstance.

Disruptions in this economy or process may be interpreted as ugly and, therefore, frequently are determined to be a nonresource. A disruption within an artificial system, which inhibits the intent of that system, is initially considered a nonresource (chaos), and that which enhances an artificial system is ultimately considered a resource (information). Similarly, within natural systems when the flow of energy is impaired or retarded, it obstructs or inhibits the prescribed pattern (Odum 1996). "Historically, the nature and value of Earth's life support systems have largely been ignored until their disruption or loss highlighted their importance" (Daily et al. 1997:1).

Rarity: From Scarcity

Scarcity is an aesthetic frequently based on a particular culture's anticipated level of an optimal abundance or quantity of a resource. Scarcity is a disruption of that continuation of an essential resource, or norm. Scarcity is defined as an insufficiency or shortness of supply; scarce—insufficient to satisfy the need or demand, not abundant (*Webster's Unabridged Dictionary of the English Language*, s.v. "Scarcity"). The golden mouse is exemplary of rarity that is based on human value, brought about by the intrigue of the animal and its habitat, *seldom* from an insufficient number of golden mice. The golden mouse is determined as exceptional or uncommon, unlike the rarity of the passenger or wild pigeon (*Ectopistes migratorius*), which was based on an anticipated abundance of individuals:

> In 1813, John James Audubon observed, along the banks of the Green River, in the State of Kentucky, "The pigeons, arriving by thousands, alighted everywhere, one above another, until solid masses, as large as hogsheads, were formed on the

branches all around. Here and there the perches gave way under the weight with a crash, and falling to the ground destroyed hundreds of the birds beneath, forcing down the dense groups with which every stick was loaded. It was a scene of uproar and confusion. I found it quite useless to speak or even to shout to those persons who were nearest to me. Even the reports of the guns were seldom heard, and I was made aware of the firing only by seeing the shooters reloading. (Vail 1920b:818)

The pigeon was abundant, making up 25–40 percent of the total bird population in the United States (three billion to five billion) when Europeans set foot on American soil. Then it became rare by the early 1900s with the last authenticated record of a capture at Sargents, Pike County, Ohio on 24 March 1900 and, finally, it became extinct with the death of Martha at age 29 on 1 September 1914.

The landscape mosaic is reflective of this type of human consideration (Forman 1997). Figure 11.14 articulates the scarcity and abundance of natural and artificial resources. (S_aS_n) represents *minimal* artificial and natural resources for human survival. Note that the golden mouse prefers disturbed systems (e.g., timbered forests), living efficiently in (S_aS_n) ecological systems. (A_aA_n)

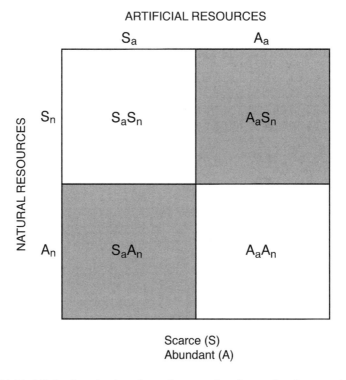

FIGURE 11.14. Minimal, optimal, and carrying capacity of natural and economic capital influence rarity (e.g., passenger pigeon [*Ectopistes migratorius*]) from scarcity (i.e., loss of a culturally anticipated abundance). After Barrett and Barrett 2001.

represents *optimal* artificial and natural resources. (A_aS_n) and (S_aA_n) represent a *carrying capacity*, respectively, with (A_aS_n) having abundant economic capital (i.e., artificial resources) with scarce or depleted natural resources, and with (S_aA_n) having scarce economic capital with abundant natural capital (nature's goods and services). Therefore, the minimal, optimal, and carrying capacity of artificial and natural systems frequently are determined by the perception and selection of natural resources by humans.

Rarity: From Biodiversity

Rarity is dependent on the abundance provided by diversity. The golden mouse is thinly distributed, few in number, and widely separated in landscape patches. With the intense scrutiny of the natural landscape in the determination of biodiversity as an essential resource for the human condition, the golden mouse might also be considered an indicator species that is uniquely reciprocal to the perception of a changing world (see Chapter 9 of this volume). Its recognition is coming less from scarcity, which represents an insufficient number of individuals or low abundance (e.g., the panda bear [*Ailuropoda melanoleuca*], the American turkey [*Meleagris gallopavo*], or the cod [*Gadus morhua*]) (Leakey and Lewin 1995, Simberloff 1994), than from rarity. Species such as the golden mouse possibly have escaped human attention as a resource or a nonresource, only to be noticed as a curiosity. Such was the case when Mr. Nuttall brought the golden mouse to the attention of Dr. Harlan (see Chapter 2 of this volume). In 1832, R. Harlan, M.D. first described the genus *Arvicola*, or *A. nuttalli*, as follows:

> The specimen under consideration is a young male, just full grown, in colour it displays a striking resemblance to the GERBILLUS *canadensis*; it was recently taken in Virginia, by Mr. Nuttall (the eminent botanist), in the vicinity of Norfolk, near the river shore, and was one of several he discovered under the bark of a hollow tree, where they had built a fine nest. (Harlan 1832:447)

Rarity, from the Latin *rarns* meaning loose, wide apart, thin, infrequent, has become synonymous with the term exceptional, extraordinary, and choice (*Webster's Unabridged Dictionary of the English Language*, s.v. "Rarity"). Rarity is a societal or individual aesthetic determined by the quality of a resource based on a level of tolerance.

Rare resources are *monitored by cultures* through inventory and investigation (Pimm and Brown 2004, Willis et al. 2004) to preserve and explore that which is essential for their survival. From this monitoring, *optimality criteria* (i.e., a set of best alternatives) are essential to guide the process of management for species survival. The situation in which nonresources are recognized as potential resources is the *optimal behavior* for management by cultures. The aesthetic of the golden mouse or its economic contribution to human survival has recent reappreciation. For example, the golden mouse enhances biodiversity. It is among relatively rare species of fauna and flora that exist parallel to humans across unique or disturbed landscapes. These landscapes allow for an *abundance of difference* (i.e., from undisturbed to habitat fragmentation). Rarity depends on the cultural perception of abundance. Whereas scarcity is distilled from dwindling numbers, rarity might be determined by the degree of difficulty in obtaining, such as the precious stone from the deep mine of Brazil or Borneo, or accessibility, such as in monitoring the ephemeral phenomenon of the lengthy annual migrations of the Monarch butterfly (*Danaus plexippus*).

Ecosystem and landscape management typically focus on physical (abiotic) factors, such as soil chemistry, rainfall, and topography, and biological processes, such as biotic diversity (genetic, species habitat), primary productivity, and stages of ecosystem development. Cultural factors, such as philosophy (i.e., perspectives on knowledge and value), need to be included within this management process (Berleant 1992).

> By introducing the notion of culture into landscape restoration, we have to broaden our conceptual and methodological scope from the natural sciences to the humanities, from strictly bioecological issues to much more complex human–ecological issues . . . exclusively discipline-oriented and mostly reductionistic scientific paradigms must be replaced by transdisciplinary concepts and methods, based on holistic [Odum and Barrett 2005] and hierarchical [O'Neill et al. 1986] systems view and its innovative approaches to wholeness and complexity . . . above all, to acknowledge the interconnected, nonlinear, mostly cybernetic and sometimes even chaotic relations between natural systems and human systems. (Naveh 1998)

Although members of each society uniquely recognize and engage "the beauty within nature," few appreciate how the *field of aesthetics* influences resource management, determines policy, and provides protection regarding natural capital—even the ecology and conservation of relatively rare microfauna such as *O. nuttalli*.

Literature Cited

Ahl, V., and T.F.H. Allen. 1996. Hierarchy theory: A vision, vocabulary, and epistemology. Columbia University Press, New York, New York.

Andreassen, H.P., S. Halle, and R.A. Ims. 1996. Optimum width of movement corridors for voles: Not too narrow, not too wide. Journal of Applied Ecology 33:63–70.

Andrews, M. 1999. Landscape and western art. Oxford University Press, New York, New York.

Audubon, J.J., and J. Bachman. 1841. Descriptions of new species of quadrupeds inhabiting North America. Pages 92–103 *in* Proceedings of the Academy of Natural Sciences of Philadelphia (1843), Vol. 1. Merrihew & Thompson, Philadelphia, Pennsylvania.

Audubon's wildlife: The quadrupeds of North America, complete and unabridged ed. (1989). Longmeadow Press, Stamford, Connecticut. (Originally published in 1846 as Viviparous quadrupeds of North America, 3 volumes. J.T. Bowen, Philadelphia, Pennsylvania).

AVI. 2007a. Poppy, rev. ed. Harper Trophy, HarperCollins Publishers, New York, New York.

AVI. 2007b. Ragweed, rev. ed. Harper Trophy, HarperCollins Publishers, New York, New York.

Barber, T.X. 1994. The human nature of birds: A scientific discovery with startling implications. Penguin Books, New York, New York.

Barbour, R.W. 1942. Nests and habitat of the golden mouse in Kentucky. Journal of Mammalogy 23:90–91.

Barrett, G.W. 1989. Viewpoint: A sustainable society. BioScience 39:754.

Barrett, G.W., and T.L. Barrett. 2001. Cemeteries as repositories of natural and cultural diversity. Conservation Biology 15:1820–1824.

Barrett, G.W., and E.P. Odum. 2000. The twenty-first century: The world at carrying capacity. BioScience 50:363–368.

Barrett, G.W., T.L. Barrett, and J.D. Peles. 1999. Managing agroecosystems as agrolandscapes: Reconnecting agricultural and urban landscapes. Pages 197–213 in W.W. Collins and C.O. Qualset, editors. Biodiversity in agroecosystems. CRC Press, Boca Raton, Florida.

Barrett, G.W., J.D. Peles, and S.J. Harper, 1995. Reflections on the use of experimental landscapes in mammalian ecology. Pages 157–174 in W.Z. Lidicker, Jr., editor. Landscape approaches in mammalian ecology and conservation. University of Minnesota Press, Minneapolis, Minnesota.

Barrett, G.W., J.D. Peles, and E.P. Odum. 1997. Transcending processes and the levels-of-organization concept. BioScience 50:363–368.

Barrett, T.L. 1997. The emperor's moth. Pages 554–561 in P.A. Gowaty, editor. Feminism and evolutionary biology: Boundaries, intersections, and frontiers. Chapman & Hall, New York, New York.

Baxley, A.M. 2005. The practical significance of taste in Kant's Critique of Judgment: Love of natural beauty as a mark of moral character. The Journal of Aesthetics and Art Criticism 63:33–45.

Berleant, A. 1992. The aesthetics of environment. Temple University Press, Philadelphia, Pennsylvania.

Bourassa, S.C. 1991. The aesthetics of landscape. Belhaven Press, London, United Kingdom.

Bull, J., and J. Farrand, Jr. 1977. The Audubon Society field guide to North American birds: Eastern region. Alfred A. Knopf, New York, New York.

Cairns, J., Jr. 1997. Global coevolution of natural systems and human societies. Revista de la Sociedad Mexicana de Historia Natural 47:217–228.

Christopher, C.C., and G.W. Barrett. 2006. Coexistence of white-footed mice, Peromycus leucopus, and golden mice, Ochrotomys nuttalli, in a southeastern forest. Journal of Mammalogy 87:102–107.

Christopher, C.C., and G.W. Barrett. 2007. Double captures of Peromyscus leucopus and Ochrotomys nuttalli. Southeastern Naturalist, In press.

Clarke, R.P. 2000. Global life systems. Rowman and Littlefield Publishers, Lanham, Maryland.

Costanza, R., R. d'Arge, R. de Groot, S. Faber, M. Grasso, B. Hannon, K. Limburg, S. Naeem, R.V. O'Neill, J. Paurelo, R.G. Raskin, P. Sutton, and M. Van Den Belt. 1997. The value of the world's ecosystem services and natural capital. Nature 387:253–260.

Covey, S.R. 1989. The 7 habits of highly effective people. Simon and Schuster, New York, New York.

Crevar, A. 2000. Endangered today, but reinvented tomorrow? Georgia Magazine 79:23.

Daily, G.C., S. Alexander, P.R. Ehrlich, L. Goulder, J. Lubchenco, P.A. Matson, H.A. Mooney, S. Postel, S.H. Schneider, D. Tilman, and G.W. Woodwell. 1997. Ecosystem services: Benefits supplied to human societies by natural ecosystems. Issues in Ecology 2. Ecological Society of America, Washington, DC.

Daily, G.C., T. Söderqvist, S. Aniyar, K. Arrow, P. Dasgupta, P.R. Ehrlich, C. Folke, A. Jansson, B. Jansson, N. Kautsky, S. Levin, J. Lubchenco, K. Mäler, D. Simpson, D. Starrett, D. Tilman, and B. Walker. 2000. The value of nature and the nature of value. Science 289:395–396.

Darwin, C. 1965. The expression of the emotions in man and animals, rev. ed. University of Chicago Press, Chicago, Illinois. (Originally published in 1872).

Doczi, G. 1994. The power of limits: Proportional harmonies in nature, art, and architecture. Shambhala Publications, Boston, Massachusetts.

Dueser, R.D., and H.H. Shugart, Jr. 1978. Microhabitats in a forest-floor small mammal fauna. Ecology 59:89–98.

Duke, D.L., M. Hebblewhite, P.C. Paquet, C. Callaghan, and M. Percy. 2001. Restoring a large-carnivore corridor in Banff National Park. Pages 261–275 *in* D.S. Maehr, R.F. Noss, and J.L. Larkin, editors. Large mammal restoration: Ecological and sociological challenges in the twenty-first century. Island Press, Washington, DC.

Farina, A. 2000. The cultural landscape as a model for the integration of ecology and economics. BioScience 50:313–320.

Forman, R.T.T. 1997. Land mosaics: The ecology of landscapes and regions. University of Cambridge, Cambridge, Massachusetts.

Gans, H.J. 1999. Popular culture and high culture. Perseus Books Group, Basic Books, New York, New York.

Goodpaster, W.W., and D.F. Hoffmeister. 1954. Life history of the golden mouse, *Peromyscus nuttalli*, in Kentucky. Journal of Mammalogy 35:16-27.

Gut, H.J., and C.E. Ray. 1963. The Pleistocene vertebrate fauna of Reddick, Florida. Quarterly Journal of the Florida Academy of Science 26:315–328.

Haeckel, E. 1869. Über entwickelungsgang und aufgabe der zoologie. Jenaische Zeitschrift für Medizin und Naturwissenschaft 5:353–370.

Hall, E.R. 1981. The mammals of North America, Vol. 2, 2nd ed. John Wiley & Sons, New York, New York.

Harlan, R. 1832. *ARVICOLA NUTTALLI*: Description of a new species of quadruped of the genus *Arvicola*, of Lacepede, or *Hypudceus*, of Illiger. Monthly American Journal of Geology and Natural Science 1:446–447.

Harris, L.D., and J. Scheck. 1991. From implications to applications: The dispersal corridor principle applied to the conservation of biological diversity. Page 309 *in* D.A. Saunders and R.R. Hobbs, editors. Nature conservation 2: The role of corridors. Surrey Beatty & Sons, Sydney, Australia.

Henderson, M.T., G. Merriam, and J.F. Wegner. 1985. Patchy environments and species survival: Chipmunks in an agricultural mosaic. Biological Conservation 31:95–105.

Huyghe. R. 1962. Art forms and society. Pages 5–12 *in* M. Heron, C. Lambert, and W. Schurmann, translators. Larousse encyclopedia of prehistoric and ancient art: Art and mankind. Prometheus Press, New York, New York. (Original from Vol. 1 of L'art et l'homme)

Jackson, L. 2000. Literature, psychoanalysis and the new sciences of mind. Longman, Pearson Education Ltd., Harlow, United Kingdom.

Jennison, C.A., L.R. Rodas, and G.W. Barrett. 2006. *Cuterebra fontinella* parasitism on *Peromyscus leucopus* and *Ochrotomys nuttalli*. Southeastern Naturalist 5:157–164.

Karlqvist, A., and U. Svedin. 1993. Introduction: From Gödel to Gaia—nature as the nonartificial. Pages 5–8 *in* H. Haken, A. Karlqvist, and U. Svedin, editors. The machine as metaphor and tool. Springer-Verlag, New York, New York.

Klee, R.V., A.C. Mahoney, C.C. Christopher, and G.W. Barrett. 2004. Riverine peninsulas: An experimental approach to homing in white-footed mice (*Peromycus leucopus*). American Midland Naturalist 151:408–413.

Kleiner, F.S., C.J. Mamiya, and R.G. Tansey. 2001. Gardner's art through the ages, 11th ed. Thomson Wadsworth, United States of America.

Kolasa, J., and N. Waltho. 1998. A hierarchical view of habitat and its relationship to species abundance. Pages 55–76 *in* D.L. Peterson and Y.T. Parker, editors. Ecological scale: Theory and applications. Columbia University Press, New York, New York.

LaPolla, V.N., and G.W. Barrett. 1993. Effects of corridor width and presence on the population dynamics of the meadow vole, *Microtus pennsylvanicus*. Landscape Ecology 8:25–37.

Leakey, R., and R. Lewin. 1995. The sixth extinction: Patterns of life and the future of humankind. Anchor Books, Random House, New York, New York.

Leopold, A. 1949. A Sand County almanac: And sketches here and there. Oxford University Press, New York, New York.

Linzey, D.W., and R.L. Packard. 1977. *Ochrotomys nuttalli*. Mammalian Species 75:1–6.

Lippard, L.R. 1999. On the beaten track. The New Press, New York, New York.

Lock, A., and C.R. Peters, editors. 1996. Handbook of human symbolic evolution. Clarendon Press, Oxford, United Kingdom.

Lubchenco, J., A.M. Olson, L.B. Brubaker, S.R. Carpenter, M.M. Holland, S.P. Hubbell, S.A. Levin, J.A. MacMahon, P.A. Matson, J.M. Melillo, H.A. Mooney, C.H. Peterson, H.R. Pulliam, L.A. Real, P.J. Regal, and P.G. Risser. 1991.

The sustainable biosphere initiative: An ecological research agenda. Ecology 72:371–412.

Maehr, D.S. 2001. Large mammal restoration: Too real to be possible? Pages 345–354 *in* D.S. Maehr, R.F. Noss, and J.L. Larkin, editors. Large mammal restoration: Ecological and sociological challenges in the twenty-first century. Island Press, Washington, DC.

Maehr, D.S., T.S. Hoctor, and L.D. Harris. 2001. The Florida panther: A flagship for regional restoration. Pages 293–312 *in* D.S. Maehr, R.F. Noss, and J.L. Larkin, editors. Large mammal restoration: Ecological and sociological challenges in the twenty-first century. Island Press, Washington, DC.

Margulis, L., and D. Sagan. 1997. Microcosms: Four billion years of microbial evolution. University of California Press, Berkeley, California.

Marzluff, J.M., and T. Angell. 2005. In the company of crows and ravens. Yale University Press, New Haven, Connecticut.

McFarland, D. 2006. Oxford Dictionary of Animal Behavior. Oxford University Press, Oxford, United Kingdom.

Merkel, A. 1998. The role of science in sustainable development. Science 281:336–337.

Metzger, C.R. 1954. Emerson and Greenough: Transcendental pioneers of an American esthetic. University of California Press, Berkeley, California.

Miller, D.A., R.E. Thill, M.A. Melchiors, T.B. Wigley, and P.A. Tappe, 2004. Small mammal communities of streamside management zones in intensively managed pine forests of Arkansas. Forest Ecology and Management 203:381–393.

Minnich, H.C. 1936. William Holmes McGuffey and his readers. American Book, New York, New York.

Mitchell, W.J. 2000. e-topia: "Urban life, Jim—but not as we know it." The MIT Press, Cambridge, Massachusetts.

Morris, J. 1967. The silk road. Horizon 9:4–23.

Mosley, L. 1990. Disney's world: A biography by Leonard Mosley. Scarborough House Publishers, Lanham, Maryland.

Murphy, S.R. 1970. Elmo Doolan and the search for the golden mouse. The Viking Press, New York, New York.

Nassauer, J.I. 1988. The aesthetics of horticulture: Neatness as a form of care. HortScience 23:973–977.

Naveh, Z. 1998. Ecological and cultural landscape restoration and the cultural evolution towards a post-industrial symbiosis between human society and nature. Restoration Ecology 6:135–143.

Noss, R.F. 2001. Introduction: Why restore large mammals? Pages 1–21 *in* D.S. Maehr, R.F. Noss, and J.L. Larkin, editors. Large mammal restoration: Ecological and sociological challenges in the twenty-first century. Island Press, Washington, DC.

Odum, E.P. 1969. The strategy of ecosystem development. Science 164:262–270.

Odum, E.P. 1984. The mesocosm. BioScience 34:558–562.

Odum, E.P., and G.W. Barrett. 2005. Fundamentals of ecology, 5th ed. Thomson Brooks/Cole, Belmont, California.

Odum, H.T. 1996. Environmental accounting, eMergy, and decision making. John Wiley & Sons, New York, New York.

Odum, H.T., and E.P. Odum. 2000. The energetic basis for valuation of ecosystem services. Ecosystems 3:21–23.

Odum, W.E., E.P. Odum, and H.T. Odum. 1995. Nature's pulsing paradigm. Estuaries 18:547–555.

O'Neill, R.V., D.L. DeAngelis, J.B. Waide, and T.F.H. Allen. 1986. A hierarchical concept of ecosystems. Monograph in Population Biology 23. Princeton University Press, Princeton, New Jersey.

Osgood, W.H. 1909. A revision of the mice of American genus *Peromyscus*. Pages 9–285 *in* United States Department of Agriculture, Bureau of Biological Survey 28-32, North American fauna 28. Government Printing Office, Washington, DC.

Packard, R.L. 1969. Taxonomic review of the golden mouse, *Ochrotomys nuttalli*. Miscellaneous Publication 51, Pages 373–406. University of Kansas Museum of Natural History, Lawrence, Kansas.

Patton, J.L., and T.C. Hsu. 1967. Chromosomes of the golden mouse, *Peromyscus* (*Ochrotomys*) *nuttalli* (Harlan). Journal of Mammalogy 48:637–639.

Perry, J., and J.G. Perry. 1994. The nature of Florida. Sandhill Crane Press, Gainesville, Florida.

Pimm, S.L., and J.H. Brown. 2004. Domains of diversity. Science 304:831–833.

Pinkham, C.F.A. 1971. *Peromyscus* and *Ochrotomys* from the Pleistocene of Reddick, Florida. Journal of Mammalogy 52:28–40.

Posey, D.A. 1999. Cultural and spiritual values of biodiversity. Page 549 *in* United Nations Environmental Programme: A Complementary Contribution to the Global Biodiversity Assessment. Intermediate Technology Publications, London, United Kingdom.

Pruett, A.L., C.C. Christopher, and G.W. Barrett. 2002. Effects of a forested peninsula on mean home range size of the golden mouse, *Ochrotomys nuttalli*, and the white-footed mouse, *Peromycus leucopus*. Georgia Journal of Science 60:201–208.

Ritvo, H. 1987. The animal estate: The English and other creatures in the Victorian Age. Harvard University Press, Cambridge, Massachusetts.

Rosenberg, D.K., B.R. Noon, and E.C. Meslow. 1997. Biological corridors: Form, function, and efficacy. BioScience 47:677–687.

Siebert, C. 2006. The animal self. New York Times Magazine, 22 January 2006.

Simberloff, D. 1994. The ecology of extinction. Acta Palaeontologica Polonica. 38:168–171.

Stack, G.J. 2005. Nietzsche's anthropic circle: Man, science, and myth. University of Rochester Press, Rochester, New York.

Swing, T.K. 1969. Kant's transcendental logic. Yale University Press, New Haven, Connecticut.

Tainter. J.A. 1988. The collapse of complex societies. Cambridge University Press, Cambridge, Massachusetts.

Taub, F.B. 1997. Unique information contributed by multispecies systems: Examples from the standardized aquatic microcosm. Ecological Applications 7:1103–1110.

The idea of nature: By R.G. Collingwood (1945). Clarendon Press, Oxford, United Kingdom.

Turner, M.G., R.H. Gardner, and R.V. O'Neill., 1995. Ecological dynamics at broad scales: Ecosystems and landscapes. BioScience Supplement S29–S35.

Vail, H.H. 1909. McGuffey's eclectic primer, rev. ed. Eclectic Educational Series. American Book, New York, New York. (Originally published in 1881)

Vail, H.H. 1920a. McGuffey's third eclectic reader, rev. ed. Eclectic Educational Series. American Book, New York, New York. (Originally published in 1879)

Vail, H.H. 1920b. McGuffey's fifth eclectic reader, rev. ed. Eclectic Educational Series. American Book, New York, New York. (Originally published in 1879)

Werner, B.T. 1999. Complexity in natural landform patterns. Science 284:102–104.

Willis, K.J., L. Gillson, and T.M. Brncic. 2004. How "virgin" is virgin rainforest. Science 304:402–403.

Wilson, D.E., and S. Ruff, editors. 1999. The Smithsonian book of North American mammals. Smithsonian Institution Press, Washington, DC.

Zonneveld, I.S. 1995. Land ecology. SPB Academic Publishing, Amsterdam, The Netherlands.

12
Future Challenges and Research Opportunities: What Do We Really Know?

Gary W. Barrett and George A. Feldhamer

To be conscious that you are ignorant is a great step to knowledge.
(Disraeli 1845:40)

As noted in Chapters 1 and 3, the golden mouse (*Ochrotomys nuttalli*) is too seldom the focal species of small mammal studies. Instead, information on the biology and ecology of the golden mouse is often gained in conjunction with studies on much more common, widespread species such as white-footed mice (*Peromyscus leucopus*) or cotton mice (*P. gossypinus*). Nonetheless, a moderate amount of information has been accumulated on the golden mouse, as described in the preceding chapters. However, just as important as defining what we know, each chapter points out the many areas in which we are clearly ignorant about interesting aspects of the life history of the golden mouse. Just as we have structured this volume on an ecological levels-of-organization approach, in this chapter we will use that same approach to summarize and address some of the future research challenges and opportunities regarding golden mice.

Research Challenges and Opportunities at the Organism/Population Level of Organization

Certain parameters of the population biology and ecology of golden mice are fairly well recognized, and some generalizations can be made. One such parameter deals with abundance. Although the local abundance of any species is dynamic spatially and temporally, as discussed in Chapter 7 of this volume, the golden mouse is probably best described as uncommon throughout its fairly extensive geographic range—considered by Schoener (1987) as "suffusive rarity." Regardless, compared to sympatric species such as white-footed mice or cotton mice, golden mice usually are less common. Also, although golden mice, throughout their fairly broad geographic range are restricted to Temperate Deciduous Forests, within this forest biome, they occur in a variety of ecosystem types ranging from uplands to lowlands.

Fewer, if any, generalizations are possible regarding the population genetics of the golden mouse. Although numerous studies have been conducted on various species and species groups of *Peromyscus*, there have been no comparable studies on golden mice. Lack of population genetics data on *O. nuttalli* is especially mystifying, given some of their behavioral characteristics, including communal nesting often observed in arboreal nests. Although one might expect less genetic diversity in populations of golden mice relative to sympatric *Peromyscus* based on relative population sizes, we encourage investigations to explore such intriguing questions as the degree of kinship of communally nesting individuals and the potential for eusociality.

Nests and nesting behavior often are a central aspect of golden mouse ecology, especially because arboreal nests are easy to locate. As a result, numerous investigators have described the physical characteristics of nests (see Linzey and Packard 1977) as well as the microhabitat variables associated with their location (Wagner et al. 2000). However, few data are available on ground nests of golden mice. Historically, ground nests have been difficult to locate, but with recent advances in miniaturization of telemetry, this is no longer a logistic problem. What is the physical structure of ground nests and their placement relative to microhabitat features? It would be fascinating to determine the thermal and physiological benefits arboreal and ground nests accrue to golden mice during different seasons and in different portions of the species range, including aspects of communal nesting. For instance, sympatric white-footed mice and deer mice (*Peromyscus maniculatus*) typically use arboreal nests at moderate temperatures such as in spring and autumn and underground nests during the heat of summer and the cold of winter (Wolff and Durr 1986; J. Wolff [personal communication]). Do golden mice follow a similar pattern of nest use and how does communal occupancy vary with season and climatic conditions? It would also be interesting to better elucidate possible gender-based differences in the use of arboreal and ground nests (Morzillo et al. 2003) and the adaptive value of such differences.

Dispersal is a critical component of population regulation and abundance, geographic range, and population genetics, as noted in Chapter 7 of this volume and elsewhere. The potential benefits and associated costs of dispersal to individual golden mice and how they affect colonization are not known. Although individuals either disperse or remain in their natal area (based on social conditions and habitat constraints discussed in Chapter 6 of this volume), effects of dispersal can be viewed at the population and community levels, as well as at higher ecosystem and landscape levels.

Given that most state and federal resource agencies are embracing the concept of multiple-use management, what is the role of golden mouse populations in the larger context of nongame (or natural heritage) programs? As discussed in Chapter 7 of this volume, whether the golden mouse is a "species of special concern" by a state resource agency might depend to a large extent on whether the state is on the periphery of the geographic range of *O. nuttalli* as well as on the amount of resources expended within the state to document the distribution and abundance of small mammal species. Regardless, in states such as Illinois, where the golden mouse is considered a threatened species, former and current forestry practices, such as logging and prescribed burning, have probably bene-fited both the density and distribution of the species. Whether trends in popula-tion abundance and distribution continue depends on the future direction of forest-management practices. This direction, in the management of natural cap-ital, is a landscape aesthetic that influences the abundance and distribution of the golden mouse (Chapter 11 of this volume).

Research Challenges and Opportunities at the Community Level of Organization

Competitive interactions among golden mice and sympatric small mammal species are one of the few areas in which a moderate amount of data is avail-able. Nonetheless, firm conclusions are problematic, and generalizations about interspecific competition in golden mice—either interference or exploitative—are more difficult to make. Although golden mice are usually less abundant than sympatric white-footed mice, Dueser and Hallett (1980) considered *O. nuttalli* a habitat specialist and the stronger competitor, as discussed in Chapter 4. Most investigators have assumed *a priori* that competition was occurring. Competitive interactions are also suggested in studies that have shown an inverse relationship between these species on different sites (Furtak-Maycroft 1991). Nonetheless, inverse relationships could also reflect selection by each species for different microhabitats. Experiments by several investiga-tors designed to demonstrate competitive release and an increase in niche width by golden mice upon removal of white-footed mice have proven equiv-ocal. Some studies have shown the expected shift (Seagle 1985), whereas oth-ers found no changes (Corgiat 1996) or a minimal effect (Christopher and Barrett 2006). Pruett et al. (2002) suggested that perhaps white-footed and golden mice were mutualistic—a conclusion reinforced by Christopher and Barrett (2007) based on double capture data in which both species occurred together unharmed in Sherman live traps. As noted by Christopher and Cameron (Chapter 4 of this volume), however, few empirical studies have determined the precise nature of interactions between golden mice and sympatric small mammal species.

Likewise, there has been speculation but little empirical data regarding the impact of predation on the ecology and community dynamics of the golden mouse. Previous investigators have assumed that because arboreal nests are usually situ-ated in a thick tangle of climbing vines, they help protect against terrestrial pred-ators from below as well as against avian predators from above (Klein and Layne 1978, Wagner et al. 2000). Such conclusions are intuitively appealing—and might be true—but are essentially speculative without rigorous experimental studies.

Research Challenges and Opportunities at the Ecosystem Level of Organization

Golden mice are an integral part of many forest ecosystems, including eastern deciduous hardwood, pinelands, riparian lowlands, floodplains, Eastern red cedar, oak-hickory, and gulf hammock forests. The undergrowth of these forests, which frequently contain an abundance of honeysuckle, greenbrier, and Chinese privet, provide the ideal habitat for golden mouse survival and reproductive success (see Chapter 5 of this volume for additional details regarding the importance of forest ecosystem structure and function on small mammal community relationships).

Two examples will demonstrate the complex relationships involving golden mice in these forest ecosystems. The flowering dogwood (*Cornus florida*) is under attack by a fatal fungal disease known as dogwood anthracnose (*Discula destructiva*). This disease has spread across nearly 4 million acres (1.6 million hectares), thereby changing the composition and aesthetics of forest ecosystems throughout the Appalachian Mountains (Bolen 1998, Chapters 5 and 11 of this volume). Rossell et al. (2001) described the impact of *D. destructiva* on the fruits of flowering dogwood, including implications of this disease for wildlife. Importantly, it is well documented that *Cornus* seeds are one of the dominant foods consumed by golden mice (Goodpaster and Hoffmeister 1954, Linzey 1968). Laura Gibbes and Luis Rodas (personal communication) have recently completed a "cafeteria-type" feeding study at the University of Georgia in which flowering dogwood fruits were the first dietary choice of both golden mice and white-footed mice. Dispersal of fruits/seeds of flowering dogwood by birds and small mammals represents a positive feedback mechanism enhancing the survival of one of the most spectacular native flowering trees (Capiello and Shadow 2005, Goodpaster and Hoffmeister 1954, Linzey 1968).

The loss of the flowering dogwood not only impacts small mammal food/energy needs but also affects the recycling of calcium within these forest ecosystems. Dogwood acts as a "calcium pump," recycling calcium between organisms and the upper layers of litter and soil. Retention and recycling of calcium in undisturbed forests is so effective that the estimated loss from the watershed is reduced to only 8 kg Ca/ha^{-1}/year^{-1}. Thus, the loss of one key species within a forest ecosystem can have cascading consequences (Figure 12.1). Unfortunately, how the loss of the flowering dogwood tree impacts golden mouse population dynamics, community relationships, and biotic diversity remain poorly understood.

A second example involves the role of golden mouse sociality (Christopher and Barrett 2007) in forest ecosystems. Chapter 4 of this volume describes the importance of competition in structuring small mammal communities, but little is known regarding the behavior and sociality of golden mice within natural and disturbed ecosystems. The social component of several species is well understood. For example, the acorn woodpecker (*Melanerpes formicivorus*) exhibits group storing and sharing of a food resource (acorns). Cavities excavated into living pines by the red-cockaded woodpecker (*Picoides borealis*) might involve the work of several birds over a period of years (Jackson 2006). Communal food hoarding and

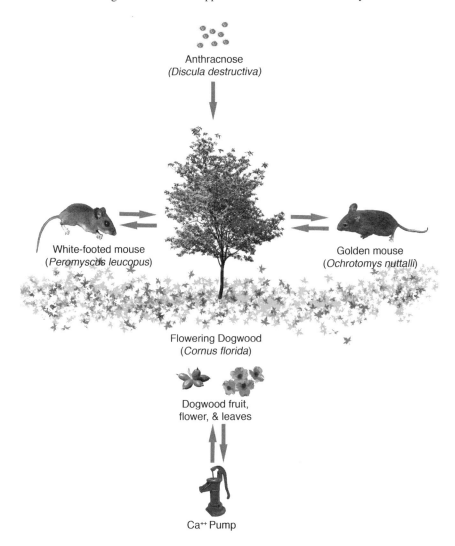

FIGURE 12.1. The relationship of white-footed mice (*P. leucopus*) and golden mice (*O. nuttalli*) to the flowering dogwood (*C. florida*) is shown. The negative effects of anthracnose (*D. destructiva*) on flowering dogwood also are depicted, which, in turn, affects the rate of calcium recovery in forest ecosystems. Graphics by Luis Rodas.

cooperative nesting have also been reported in taiga voles (*Microtus xanthognathus* [Wolff 1980]) and several species of *Peromyscus* (e.g., Wolff 1989, 1994). Extended families and communal and cooperative nest sharing also occur in montane voles (*Microtus montanus* [Jannett 1978]), prairie voles (*M. ochrogaster* [Getz et al. 1993]), and others (Solomon and Getz 1997). In general, communal sharing of winter nests often occurs among nonrelatives, whereas cooperative breeding typically involves related females (e.g., Solomon and French 1997). Do golden mice share in the building of elaborate globular nests or in the construction of even larger shelter/communal nests? Is it possible, because of the large number (six to eight) of golden mice sometimes found in a single nest, that this species exhibits *eusociality* characterized by cooperative division of labor, caring for young, and overlap of two generations functioning to contribute to group reproductive success, as has been demonstrated in *Peromyscus* (Wolff 1994)? Further, what is the social relationship of *O. nuttalli* to *P. leucopus* when sharing the same habitat and even livetrapped simultaneously (Christopher and Barrett 2007)? Under some circumstances, could this relationship be more mutualistic than competitive? Numbers of *Peromyscus* fluctuate dramatically in response to cyclic production of acorns and other mast (Wolff 1996). Whether this fluctuating resource and corresponding density changes of potential competitors affect golden mice is not known. These questions await further study and provide exciting avenues for future investigations across levels of organization.

Research Challenges and Opportunities at the Landscape Level of Organization

Barrett and Peles (1999) described how small mammals are model species to address questions at the landscape scale. It should be noted, however, that *O. nuttalli* is missing from the rich diversity of small mammal species investigated at this scale. Chapter 6 of this volume attempts to summarize what little is known regarding the role of landscape elements such as patch quality, edge habitat, landscape corridors, and matrix habitat as related to the golden mouse. The landscape and its relationship to golden mice is likely the greatest challenge (and poses the most exciting opportunities) regarding an increased understanding of this most unique small mammal species.

For example, as shown in Figure 6.4, Chapter 6 of this volume, do golden mice use either disturbance corridors (e.g., power-line corridors) or resource corridors (e.g., riparian stream habitat) as means of dispersal? Ironically, compared to numerous other vertebrate species (especially interior species), golden mice benefit from landscape fragmentation, including timber harvest, tornadoes, or hurricanes. However, the question remains: What is the optimum disturbance necessary to optimize golden mice population abundance and survivorship? Habitat fragmentation leads to increased edge habitat. We consider the golden mouse an edge species because of its use of this habitat type. Does this increase in edge habitat provide corridors for dispersal as well as optimum habitat for nesting and reproduction? Although corridors have been established at the landscape scale to investigate interpatch movements of meadow voles (*Microtus pennsylvanicus*), cotton mice, old-field mice (*P. polionotus*), and cotton rats (*Sigmodon hispidus*) (LaPolla and Barrett 1993, Mabry and Barrett 2002), no studies have been conducted to investigate corridor use or interpatch movements of golden mice.

Abundant landscape-level information exists on mammalian species such as the southern flying squirrel (*Glaucomys volans*) and the red fox (*Vulpes vulpes*). There exists a paucity of information, however, focusing on the charismatic golden mouse at this scale. This information will become increasingly vital if ecologists, land-use managers, and policy makers are to better understand ecological processes and regulatory mechanisms—including dispersal behavior, population genetics, community interactions, habitat architecture, and landscape fragmentation—across levels of organization. Future landscape-level investigations focusing on golden mice should provide new frontiers of investigation.

Concluding Remarks

Throughout this volume we have provided information on the golden mouse across levels of ecological organization: organism, population, community, ecosystem, and landscape. We trust that readers will appreciate and be amazed at the unusual habitat selection, bioenergetics, nesting behavior, sociality, community relationships, and use of landscape elements exhibited by the golden mouse—and be cognizant of the array of interesting questions yet to be addressed at each level of organization. The golden mouse is truly a unique small mammal species.

In addition to preparing a book that will appeal to readers, we also challenge undergraduate and graduate students to design and conduct experiments that will provide more insight into golden mice. Not only do golden mice exhibit charismatic appeal, but understanding of their niche in forest ecosystems should provide ecologists, mammalogists, resource managers, and policy makers with vital information necessary for the conservation and sustained management of this unique species at greater temporal and spatial scales.

Literature Cited

Barrett, G.W., and J.D. Peles. 1999. Small mammal ecology: A landscape perspective. Pages 1–10 *in* G.W. Barrett and J.D. Peles, editors. Landscape ecology of small mammals. Springer-Verlag, New York, New York.

Bolen, E.G. 1998. Ecology of North America. John Wiley & Sons, New York, New York.

Capiello, P., and D. Shadow. 2005. Dogwoods. Timber Press, Portland, Oregon.

Christopher, C.C., and G.W. Barrett. 2006. Coexistence of white-footed mice (*Peromyscus leucopus*) and golden mice (*Ochrotomys nuttalli*) in a southeastern forest. Journal of Mammalogy 87:102–107.

Christopher, C.C., and G.W. Barrett. 2007. Double captures of *Peromyscus leucopus* and *Ochrotomys nuttalli*. Southeastern Naturalist, In press.

Corgiat, D.A. 1996. Golden mouse microhabitat preference: Is there interspecific competition with white-footed mice in southern Illinois? MS thesis, Southern Illinois University, Carbondale, Illinois.

Disraeli, B., Earl of Beaconsfield. 1881. Sybil, or, The two nations. *Hugenden edition,* Book 1, Chapter 5. Page 40. Longmans, Green, and Company, London, United Kingdom. (Originally published in 1845, H. Colburn, London, United Kingdom).

Dueser, R.D., and J.G. Hallett. 1980. Competition and habitat selection in a forest-floor small mammal fauna. Oikos 35:293–297.

Furtak-Maycroft, K.A. 1991. An empirically-based habitat suitability index model for golden mice, *Ochrotomys nuttalli*, on pine stands in the Shawnee National Forest. MS thesis, Southern Illinois University, Carbondale, Illinois.

Getz, L.L., B. McGuire, T. Pizzuto, J.E. Hofmann, and B. Frase. 1993. Social organization of the prairie vole (*Microtus ochrogaster*). Journal of Mammalogy 74:44–58.

Goodpaster, W.W., and D.F. Hoffmeister. 1954. Life history of the golden mouse, *Peromyscus nuttalli*, in Kentucky. Journal of Mammalogy 35:16–27.

Jackson, J.A. 2006. In search of the ivory-billed woodpecker. Smithsonian Books/Collins, New York, New York.

Jannett, F.J., Jr. 1978. Density-dependent formation of extended maternal families of montane vole, *Microtus montanus nanus*. Behavioral Ecology and Sociobiology 3:245–263.

Klein, H.G., and J.N. Layne. 1978. Nesting behavior of four species of mice. Journal of Mammalogy 59:103–108.

LaPolla, V.N., and G.W. Barrett. 1993. Effects of corridor width and presence on the population dynamics of the meadow vole, *Microtus pennsylvanicus*. Landscape Ecology 8:25–97.

Linzey, D.W. 1968. An ecological study of the golden mouse, *Ochrotomys nuttalli*, in the Great Smoky Mountains National Park. American Midland Naturalist 79:320–345.

Linzey, D.W., and R.L. Packard. 1977. *Ochrotomys nuttalli*. Mammalian Species 75:1–6.

Mabry, K.E., and G.W. Barrett. 2002. Effects of corridors on home range size and interpatch movements of three small mammal species. Landscape Ecology 17:629–636.

Morzillo, A.T., G.A. Feldhamer, and M.C. Nicholson. 2003. Home range and nest use of the golden mouse (*Ochrotomys nuttalli*) in southern Illinois. Journal of Mammalogy 84:553–560.

Pruett, A.L., C.C. Christopher, and G.W. Barrett. 2002. Effects of a forested riparian peninsula on mean home range size of the golden mouse (*Ochrotomys nuttalli*) and the white-footed mouse (*Peromyscus leucopus*). Georgia Journal of Science 60:201–208.

Rossell, I.M., C.R. Rossell, K.J. Hining, and R.L. Anderson. 2001. Impacts of dogwood anthracnose, *Discula destructiva* Redlin, on the fruits of flowering dogwood, *Cornus florida* L.: Implications for wildlife. American Midland Naturalist 146:379–387.

Schoener, T.W. 1987. The geographical distribution of rarity. Oecologia 74:161–173.

Seagle, S.W. 1985. Competition and coexistence of small mammals in an east Tennessee pine plantation. American Midland Naturalist 114:272–282.

Solomon, N.G., and J.A. French. 1997. Cooperative breeding in mammals. Cambridge University Press, Cambridge, United Kingdom.

Solomon, N.G., and L.L. Getz. 1997. Examination of alternative hypotheses for cooperative breeding in rodents. Pages 199–230 *in* N.G. Solomon and J.A. French, editors. Cooperative breeding in mammals. Cambridge University Press, Cambridge, United Kingdom.

Wagner, D.M., G.A. Feldhamer, and J.A. Newman. 2000. Microhabitat selection by golden mice (*Ochrotomys nuttalli*) at arboreal nest sites. American Midland Naturalist 144:220–225.

Wolff, J.O. 1980. Social organization of the taiga vole (*Microtus xanthognathus*). The Biologist 62:34–45.

Wolff, J.O. 1989. Social behavior. Pages 271–291 *in* G.L. Kirkland, Jr. and J.N. Layne, editors. Advances in the study of *Peromyscus* (Rodentia). Texas Tech University Press, Lubbock, Texas.

Wolff, J.O. 1994. Reproductive success of solitarily and communally nesting white-footed mice and deer mice. Behavioral Ecology 5:206–209.

Wolff, J.O. 1996. Population fluctuations of mast-eating rodents are correlated with production of acorns. Journal of Mammalogy 77:850–856.

Wolff, J.O., and D.S. Durr. 1986. Winter nesting behavior of *Peromyscus leucopus* and *Peromyscus maniculatus*. Journal of Mammalogy 67:409–411.

Index

Printed in the United States of America